Ian Stewart is Professor of Mathematics at Warwick University. An active research mathematician, he is also a world-renowned popularizer of mathematics. He has received numerous awards and honours for his contributions to not only mathematics itself but also to the public understanding of mathematics and related areas of science.

He has written many books including *Math Hysteria*, *From Here to Infinity*, *Letters to a Young Mathematician*, three books in the *The Science of Discworld* series (with Terry Pratchett and Jack Cohen), *Flatterland*, *Nature's Numbers*, and *Does God Play Dice?*

# How to Cut a Cake

## And other mathematical conundrums

### IAN STEWART

OXFORD
UNIVERSITY PRESS

# OXFORD
UNIVERSITY PRESS

Great Clarendon Street, Oxford OX2 6DP

Oxford University Press is a department of the University of Oxford.
It furthers the University's objective of excellence in research, scholarship,
and education by publishing worldwide in

Oxford New York ·

Auckland Cape Town Dar es Salaam Hong Kong Karachi
Kuala Lumpur Madrid Melbourne Mexico City Nairobi
New Delhi Shanghai Taipei Toronto

With offices in

Argentina Austria Brazil Chile Czech Republic France Greece
Guatemala Hungary Italy Japan Poland Portugal Singapore
South Korea Switzerland Thailand Turkey Ukraine Vietnam

Oxford is a registered trade mark of Oxford University Press
in the UK and in certain other countries

Published in the United States by
Oxford University Press Inc., New York

British Library Cataloguing in Publication Data

Data available

Library of Congress Cataloging in Publication Data

Data available

Typeset by RefineCatch Limited, Bungay, Suffolk
Printed in Great Britain
on acid-free paper by
Clays Ltd., St Ives plc

ISBN 0-19-920590-6 (Pbk.)   978-0-19-920590-5 (Pbk.)

1 3 5 7 9 10 8 6 4 2

# Contents

# Preface

**O**ccasionally, when I am feeling unusually relaxed and my mind starts to wander, I wonder what the world would be like if everyone enjoyed mathematics as much as I do. Television news would lead with the latest theorems in algebraic topology instead of tawdry political scandals, teenagers would download Top of the Theorems to their iPods, and calypso singers (remember them?) would strum their guitars to the tune of 'Lemma Three' ... Which reminds me that the folk-singer Stan Kelly (now Stan Kelly-Bootle, look him up on Google) actually wrote just such a song, back in the late 1960s when he was studying for a mathematics MSc at the University of Warwick. It began:

> Lemma three, very pretty, and the converse pretty too,
> But only God and Fermat know which one of them is true.

Anyway, I have generally viewed maths as a source of inspiration and enjoyment. I am aware that in most people what it inspires is sheer terror, not fun, but I find it impossible to share that view. Rationally, I understand some of the reasons for the widespread fear of mathematics: there is nothing worse than a subject that absolutely demands accuracy and precision when you are hoping to talk your way out of trouble with a couple of buzzwords and a large dose of effrontery. But emotionally, I find it hard to understand why a subject so vital to the world we inhabit, with such a long and absorbing history, littered with the most brilliant insights ever made by human beings, can fail to intrigue and fascinate.

On the other hand, bird-watchers also find it hard to appreciate why the rest of the world does not share their passion for ticking off entries on lists. 'My god, is that the mating plumage of the lesser-crested nitwit? The last one recorded in Britain was sighted on the Isle of Skye in 1843, and that one was partially hidden behind a – oh, no, it's just a starling with mud on its tail.' No offence intended – I collect rocks. 'Wow! Genuine Aswan granite!' Our house is filling up with pieces of the planet.

It probably doesn't help that what most people mean by the word 'mathematics' is routine arithmetic. That's fun, in a nerdy sort of way, if you can do it. It's horrible if you can't. Moreover, it is very difficult to have fun with something – be it maths or bird-watching – if somebody is standing over you with a big red pen in hand, just waiting for you to make some minor slip so that they can jump in and scrawl all over it. (I mean this metaphorically. It used to be literal.) After all, what's a decimal point or two between friends? But somewhere in the gap between the National Curriculum and Young Henry's grasp of it, a lot of the fun of mathematics seems to have gone the way of the Dodo. Which is a pity.

I'm not claiming that *How to Cut a Cake* will have a dramatic effect on the mathematical abilities of the general public, though I suppose it might. (In which direction . . . ah, that's another matter.) What I'm trying to do here is mostly to preach to the converted. This is a book for the fans, for the enthusiasts, for the people who actively *like* mathematics, and who retain a sufficiently youthful mind that they can gain a lot of pleasure through play. The air of frivolity is reinforced by Spike Gerrell's delightful cartoons, which capture the spirit of the discussion perfectly.

The intent, however, is deadly serious.

I actually wanted to call the book *Weapons of Math Distraction*, which to my mind had exactly that balance of seriousness and frivolity, so I should probably be grateful for the marketing department's veto. But with its practical cake-oriented title there is a danger that some of you may be thinking of buying this book to obtain some serious culinary instruction. Whence this disclaimer: this is a book about puzzles and games of a mathematical nature, not about cookery. The cake is actually a Borel measure space.

Heavily disguised as . . . a cake. And what mathematics teaches us is not how to cook it, but how to divide it fairly among any number of people. And – which is much harder – without creating envy. Cake-cutting provides a simple introduction to mathematical theories of sharing resources. Like most introductory mathematics, it is what the professionals like to call a 'toy model', drastically simplified from anything in the real world. But it gets you thinking about some key issues. For example, it makes it obvious that it is *easier* to split resources among several competing groups, in a manner that they all consider fair, if they value them differently.

Like its predecessors *Game, Set and Math, Another Fine Math You've Got Me Into,* and *Math Hysteria* (the latter, like this one, published by Oxford University Press), the present volume derives from a series of columns on mathematical games that I wrote for *Scientific American* and its foreign-language translations between 1987 and 2001. The columns have been lightly edited, all known mistakes have been corrected, an unknown number of *new* mistakes have been introduced, and readers' comments have been inserted where appropriate under the heading 'Feedback'. I have restored some material that did not appear in the magazine versions because of space limitations, so this is a kind of 'director's cut', so to speak. The topics range from graphs to probability, from logic to minimal surfaces, from topology to quasi-crystals. And cake apportionment, of course. They have been selected primarily for amusement value, not for significance, so please don't imagine that the content is fully representative of current activity at the research frontiers.

It does however, *reflect* current activity at the research frontiers. The burning issue of cutting a cake is part of a long tradition in mathematics – it goes back at least 3500 years to ancient Babylon – of posing serious questions in frivolous settings. So when, as here, you read about 'why phone cords get tangled', the topic is not solely useful for tidying up the rat's nest of wires that typically attaches your telephone to its handpiece. The best mathematics has a curious kind of universality, so that ideas derived from some simple problem turn out to illuminate a lot of others. In the real world, many things twist and turn: phone cords, plant tendrils, DNA molecules, underwater communication cables. These four applications of the mathematics of twists and . turns differ widely in many essential respects: you would be understandably upset if the telephone engineer took away your phone cable and replaced it with a length of bindweed. But they also overlap in one useful respect: the same simple mathematical model sheds light on them all. It may not answer every question you would like to answer, and it may ignore some important practical issues, but once a simple model has opened the door to mathematical analysis, then more complex, more detailed models can be developed on that basis.

My aim here is to use a mixture of abstract thought and the real world to motivate various mathematical ideas. The payoff, for me, does not lie in obtaining practical solutions to real world problems. The main payoff is new mathematics. It's not possible to develop a major application of mathematics in a few pages, but it is possible, for anyone who has enough imagination, to appreciate how a mathematical idea derived in one setting can unexpectedly apply in a different one. Perhaps the best example in this book is the connection between 'empires' and electronic circuits. Here a strange and artificial puzzle about colouring maps of territories on the Earth and the Moon (Chapter 9) makes useful practical inroads into the important question of testing electronic circuit boards for flaws (Chapter 10). The point is that the mathematicians first stumbled across the central idea in a frivolous context (though not *quite* as frivolous as the version presented here) and only then did it become apparent that it had serious applications.

It can work the other way. Chapter 15 is inspired by the remarkable behaviour of some species of Asian firefly, in which the males synchronize their flashes – probably to improve their collective ability to attract females, though not their individual abilities. How do the flashes become synchronized? Here the serious problem came first, the mathematics addressed that problem and provided at least part of its solution, and only later did it become clear that the same mathematics can be used for many other questions about synchronization. My approach turns the whole thing into a board game that you can play. As a twist in the tale: some deceptively simple questions about that game still remain unanswered. In some ways we understand the real application better than the simple model.

With very few exceptions, each chapter stands on its own. You can dip in anywhere, and if you get stuck, for whatever reason, you can abandon that chapter and try another one. What you will gain – I claim – is an increased understanding of just how broad the subject of mathematics is, of how much further it goes than anything that is ever taught at school, of its astonishingly wide range of applications, and of the surprising cross-connections that bind the entire subject together into a single, amazingly powerful package. All achieved by solving puzzles and playing games.

And, more importantly, by stretching your mind.

Never underestimate the power of play.

IAN STEWART
*Coventry, April 2006*

# Figure Permissions

# Your Half's Bigger than My Half!

If two people want to share a cake, with no arguments, then the time-honoured solution is 'I cut, you choose'. The problem becomes surprisingly tricky with more than two people, and the more people there are, the trickier it becomes. Unless you use a slowly moving knife to cut through the difficulties ... and the cake.

A **big man and a small man** were sitting in the restaurant car of a train, and both ordered fish. When the waiter brought the food, there was one big fish and one small one. The big man, served first, promptly took the big fish; the small man complained that this was extremely impolite.

'What would you have done if *you'd* been offered first choice, then?' asked the big man, in a tone of annoyance.

'I would have been polite and taken the small fish,' said the small man smugly.

'Well, that's what you've got!' replied the big man.

As this ancient joke illustrates, different people place different values on things under different circumstances, and some folk are very hard to please. For the past fifty years, mathematicians have grappled with problems of fair division – usually formulated in terms of a cake rather than fish – and there is now an extensive and surprisingly deep theory. Jack Robertson and William Webb's fascinating book *Cake Cutting Algorithms* (see Further Reading for details) surveys the entire field. In this chapter and Chapter 14 we'll take a look at some of the ideas that have emerged from the deceptively simple question of dividing a cake so that everybody is happy with their share.

The simplest case involves just two people, who – to reiterate – wish to share a cake so that each is satisfied that they have a fair share. 'Fair' here means 'more than half by *my* valuation', and the recipients may disagree on the value of any given bit of cake. For example, Alice may like cherries while Bob prefers icing. One of the more curious insights that has emerged from the theory of cake cutting is that it is *easier* to divide the cake when the recipients disagree on what parts of it are worth. You can see this makes sense here, because we can give Bob the icing and Alice the cherries and we're well on the way to satisfying both of them. If they both wanted icing, the problem would be harder.

Not that it's terribly hard when there are two players. The solution 'Alice cuts, Bob chooses' has been traced back 2800 years! Both players find this fair in the sense that they have no right to complain about the end result. If Alice dislikes the piece that Bob leaves, it's her own fault for not being more careful to make equal cuts (according to her valuation). If Bob doesn't like his piece, he made the wrong choice.

The whole area began to get interesting when people looked at what happens with three players. Tom, Dick, and Harry want to divide a cake so that each is satisfied he's got at least one third of it, according to his own private valuation. In all such matters, by the way, the cake is assumed to be infinitely divisible, although much of the theory works if the cake has valuable 'atoms' – single points to which at least one recipient attaches a non-zero value. For simplicity, though, I'll assume there are no atoms. Robertson and Webb approach this variant by analysing a plausible but incorrect answer, which goes like this.

STEP 1: Tom cuts the cake into two pieces $X$ and $W$, where he thinks that $X$ is worth 1/3 and $W$ is worth 2/3.

STEP 2: Dick cuts $W$ into two pieces $Y$ and $Z$, which he thinks are each worth 1/2 of $W$.

STEP 3: Harry chooses whichever of $X$, $Y$, and $Z$ he prefers. Then Tom chooses from the two pieces left. Dick gets the last piece.

Is this algorithm fair?

It's clear that Harry will be satisfied, because he has first pick. Tom is also satisfied, for slightly more complex reasons. If Harry picks $X$, then

Tom can pick whichever of $Y$ and $Z$ he considers more valuable (or either if they are equal in his eyes). Since he thinks they are worth 2/3 in total, he must think at least one of them is worth 1/3. On the other hand, if Harry chooses $Y$ or $Z$, then Tom can choose $X$.

However, Dick may not be so happy with the result. If he disagrees with Tom about the first cut, then he might think $W$ is worth less than 1/3, meaning that the only piece that will satisfy him is $X$. But Harry could choose $Y$, say, and Tom $X$, so Dick has to take $Z$ – which he doesn't want.

The above algorithm, then, is not fair. The first correct solution to fair three-person division was given in 1944 by Hugo Steinhaus, one of a group of Polish mathematicians who met regularly in a café in Lvov. His method involves a technique called 'trimming'.

STEP 1: Tom cuts the cake into two pieces $X$ and $W$, where he thinks that $X$ is worth 1/3 and $W$ is worth 2/3.

STEP 2: He passes $X$ to Dick and asks him to trim it so that Dick values it at 1/3, if he thinks it's worth more than that, and to leave it alone if not. Call the resulting piece $X^*$: this is either $X$ or smaller.

STEP 3: Dick passes $X^*$ to Harry, who can either agree to take it, or not.

STEP 4: (a) If Harry accepts $X^*$ then Tom and Dick pile the rest of the cake – $W$ plus any trimmings from $X$ – in a heap, and treat this as a single (messy) cake. They play 'I cut you choose' on that. (b) If Harry does not accept $X^*$ *and* Dick has trimmed $X$, then Dick takes $X^*$, and Tom and Harry play 'I cut you choose' on the rest. (c) If Harry does not accept $X^*$ *and* Dick has not trimmed $X$, then Tom takes $X$, and Dick and Harry play 'I cut you choose' on the rest.

That's one answer – I'll leave it to you to verify the logic. Basically, anyone who isn't satisfied with what he gets must have made a bad choice, or a poorly judged cut, at an earlier stage, in which case he has only himself to blame.

In 1961 Leonard Dubins and Edwin Spanier proposed a rather different solution involving a moving knife. Sit the cake on a table, and arrange for a knife to move smoothly and gradually across it, starting completely to its left. At a given instant, let $L$ be the part to the left of the knife. Tom, Dick, and Harry are all told to shout 'Stop!' as soon as the value of $L$, in their opinion, becomes 1/3. The first to shout gets $L$, and the other two divide the rest either by 'I cut you choose' or by moving the knife again and shouting as soon as the perceived value reaches 1/2. (What should they do if two players shout simultaneously? Think about it.)

The great feature of this method is that it extends readily to $n$ recipients. Move the knife across, and tell everyone to shout as soon as $L$ reaches $1/n$ in their estimation. The first person to shout gets $L$, and the remaining $n - 1$ players repeat the process on the remaining cake, only of course they now shout when the perceived value reaches $1/(n - 1)$ . . . and so on.

I have never been terribly happy about moving-knife algorithms – I think because of the time-lag involved in the players' reactions. Perhaps the best way to get round this quibble is to move the knife slowly. *Very* slowly. Or, equivalently, to assume that all of the players have super-fast reactions.

Let's call the first kind of answer a 'fixed knife' algorithm, the second a 'moving knife' algorithm. There is a fixed knife algorithm for three-person division that also extends readily to $n$ people. Tom is sitting on his own, staring at 'his' cake, when Dick shows up and asks for a share. So Tom cuts what he thinks are halves and Dick chooses a piece. Before they can eat anything, Harry arrives and demands a fair share too. Tom and Dick independently cut their pieces into three parts, each of which they consider to be of equal value. Harry chooses one of Tom's pieces and one of Dick's. It's not hard to see why this 'successive pairs' algorithm works, and the extension to any number of people is relatively straightforward. The trimming method can also be extended to $n$ people by offering everyone round the table a chance to trim a piece if they are willing to accept the result, and insisting that they do if nobody else wants to trim it further.

When the number of people is large, the successive pairs algorithm requires a very large number of cuts. Which method requires the fewest cuts? The moving knife method uses $n - 1$ cuts to get its $n$ pieces, and that's as small as you can get. But the fixed knife methods don't succumb as readily. With $n$ people, a generalization of the trimming algorithm uses $(n^2 - n)/2$ cuts. The successive pairs algorithm uses $n! - 1$, where $n! = n(n - 1)(n - 2) \ldots 3.2.1$ is the *factorial* of $n$. This is bigger than the number of cuts used in the trimming algorithm (except when $n = 2$).

However, trimming is not the best method. The more efficient 'divide and conquer' algorithm works roughly like this: try to divide the cake using one cut so that roughly half the people would be happy to have a fair share of one piece, while the rest would be happy to have a fair share of the other piece. Then repeat the same idea on the two separate subcakes. The number of cuts needed here is about $n \log_2 n$. The exact formula is $nk - 2^k + 1$ where $k$ is the unique integer such that $2^{k-1} < n \le 2^k$. It is conjectured that this is about as good as you can get.

These ideas could eventually go beyond mere recreation. There are many situations in real life where it is important to divide assets in a manner that seems fair to all recipients. Negotiations over territory and commercial interests are examples. In principle the kind of method that solves the cake-cutting problem can be applied to such situations. Indeed, when for administrative purposes Germany was divided among the Allies (the USA, the UK, France) and Russia, the first attempt created leftovers (Berlin) and then Berlin had to be divided as a separate step, so negotiators intuitively apply similar methods. Something rather similar is causing problems in Israeli–Palestinian relations right now, with Jerusalem as the main 'leftovers' and the

West Bank as another bone of contention. Might the mathematics of fair allocation assist the negotiations? It would be nice to think we lived in a world that was sufficiently rational for such an approach, but politics seldom works that way. In particular, people's valuations of things tend to change *after* tentative agreements have been reached, in which case what we've just discussed won't work.

Still, it could be worth giving rational methods a try.

## FEEDBACK

I received a lot of correspondence on cake-cutting algorithms, ranging from simplifications of the methods I discussed to substantial new research articles. Some readers attempted to dispel my vague feelings of disquiet about 'moving knife' algorithms. My worry was the element of reaction time. The suggestion for avoiding this problem – slightly refined by some to-and-fro correspondence – was that in place of the moving knife, the players should make marks on the cake (or a scale model). First, choose a direction (say north–south) and ask each of the $n$ players, in turn, to mark the cake with a north–south line at the westmost position for which they are willing to accept the cake to the west of the mark. (That is, where they estimate the 'worth' of the left-hand slice to be $1/n$.) Whoever's mark is furthest west cuts off that bit and exits the game. Now continue in the same general manner. The ordering of the cuts in the west–east direction substitutes for the timing, and the same idea can be used for all moving knife methods.

It looked as though my reservations about moving knife algorithms were unjustified. But soon after, Steven Brams of New York University, an expert on such matters, wrote to point out that my original worries are not so easily dismissed. In particular Brams, Alan D. Taylor, and William S. Zwicker analyse moving-knife schemes in two articles listed in Further Reading. Their second paper exhibits a moving-knife procedure for an envy-free allocation among four players, that needs at most 11 cuts.

However, no discrete procedure with a *bounded* number of cuts (however large) is known for four players, and such schemes probably don't exist. Certainly their scheme cannot be made discrete by the use of notional 'marks' on the cake. So the reduction of moving-knife schemes to 'marks' works in some cases – but not all.

# Repealing the Law of Averages

According to a folk belief often called the 'law of averages', random events should even out in the long run. So should you bet on lottery numbers that have *not* come up as often as the others? Probability theory provides a resounding 'no'. Nevertheless, there is a sense in which random events really do even out in the long run. It just won't help you win the lottery.

**S**uppose that I keep tossing a fair coin – one for which 'heads' and 'tails' are equally likely, each having probability 1/2 – and maintain a running count of how many times each turns up. How should I expect these numbers to behave? For instance, if at some stage 'heads' gets well ahead of tails – say I have tossed 100 more heads than tails – is there any tendency for tails to 'catch up' in future tosses?

People often talk of an alleged 'law of averages', based on the intuitive feeling that tosses of a fair coin ought to even up eventually. Some even believe that the probability of a tail must increase in such circumstances – often expressed as the idea that tails become 'more likely'. Others assert that coins have no memory – so the probability of heads or tails always remains 1/2 – and deduce that there is no tendency whatsoever for the numbers to even out.

Which view is right?

The same issues arise in many different circumstances. Newspapers publish tables of how often various numbers have come up in lotteries. Should such tables influence your choices? If big earthquakes normally happen in some region every fifty years, on average, and one hasn't occurred for sixty, is it 'overdue'? If aeroplane crashes happen on average once every four months, and three months have passed without one, should you expect one soon?

In all cases, the answer is 'no' – though I'm open to debate as regards earthquakes, because the absence of a big quake can often be evidence for a big build-up of stress along a fault line. The random processes involved – or, more accurately, the standard mathematical models of those processes – have no 'memory'.

This, however, is not the end of the story. A lot depends on what you mean by 'catch up'. A long run of heads does not affect the *probability* of getting a tail later on, but there is still a sense in which coin tosses tend to even out in the long run. After a run of, say, 100 more heads than tails, the probability that *at some stage* the numbers will even up again is 1. Normally, a probability of 1 means 'certain' and a probability of 0 means 'impossible', but in this case we are working with a potentially infinite list of tosses, so mathematicians prefer to say 'almost certain' and 'almost impossible'. For practical purposes you can forget the 'almost'.

The same statement applies no matter what the initial imbalance is. Even if heads are a quadrillion throws ahead, tails will still 'almost certainly' catch up if you keep tossing long enough. If you're worried that this somehow conflicts with the 'no memory' point, I hasten to add that there is also a sense in which coin tosses do *not* have a tendency to even out in the long run! For example, after a run of 100 more heads than tails, the probability that the cumulative number of heads eventually gets at least a million ahead of tails is also 1.

To see how counter-intuitive these questions are, suppose that instead of tossing a coin I roll a die (the singular of 'dice', which is actually a plural word). Count how many times each face, 1 to 6, turns up. Assume each face has probability 1/6, equally likely. When I start, the cumulative numbers of occurrences of each face are equal – all zero. Typically, after a few throws, those numbers start to differ. Indeed, it takes at least six throws before there is any chance of them evening out again, at one of each. What is the probability that however long I keep throwing the die, the six numbers *at some stage* even out again? Unlike heads and tails for a coin, this probability is *not* 1. In fact, it's less than 0.35; for the exact value, see 'Feedback' below. By appealing to some standard theorems in probability, I can easily prove to you that it's certainly not 1.

Why does a die behave differently from a coin? Before answering that, we have to take a close look at coin-tossing. A single toss of a coin is called a 'trial', and we are interested in a whole series of trials, possibly going on forever. I tossed a coin 20 times, getting the result TTTTHTHHHHHHTTTHTTTH. Here there are 11 T's and 9 H's. Is this reasonable?

The answer to such questions is given by a theorem in probability known as the Law of Large Numbers. It states that the frequencies with which events occur should, in the long run, be very close to their probabilities. Since tossing H with a fair coin has probability 1/2, by the definition of 'fair', the Law of Large Numbers tells us that 'in the long run' roughly 50% of all tosses should be heads. The same goes for T.

Similarly, with a fair die, 'in the long run' roughly 16.7% (one sixth) of all rolls should produce any given result: 1, 2, 3, 4, 5, or 6. And so on.

In my sequence of 20 coin tosses, the frequencies are $11/20 = 0.55$ and $9/20 = 0.45$, which is close to 0.5, but not equal to it. You may feel that my sequence doesn't look random *enough*. You'd probably be much happier with something like HTHHTTHTTHTHHTHTHHTT, with frequencies $10/20 = 0.5$ for H and $10/20 = 0.5$ for T. As well as getting the numbers spot on, the second sequence looks more random. But it isn't.

What makes the first sequence look non-random is that there are long repetitions of the same event, such as TTTT and HHHHHH. The second sequence lacks such repetitions, so we think it looks more random. But our intuition about what randomness looks like is misleading: random sequences *should* contain repetitions! For example, in successive blocks of four events, like this:

```
T T T T H T H H H H H H T T T H T T T H
T T T T
  T T T H
    T T H T
      T H T H
```

and so on, the sequence TTTT should occur about one time in 16. I'll explain why in a moment, but let's follow up the consequences first. For my first sequence above, TTTT occurs one time in 17 – pretty much spot on! Agreed, HHHHHH should occur only once in 64 times, on average, but it occurs once out of just 15 blocks of length 6 in my sequence – but I didn't throw my coin enough times to see whether it came up again later. *Something* has to come up, and HHHHHH is just as likely as HTHTHT or HHTHTT.

Random sequences often show occasional patterns and clumps. Don't be surprised by these: they are *not* signs that the process is non-random . . . unless the coin goes HHHHHHHHHHHH . . . for a long time, in which case the shrewd guess is that it is double-headed.

Suppose you toss four fair coins in a row. What can happen? Figure 1 summarizes the possible results. The first toss is either H or T

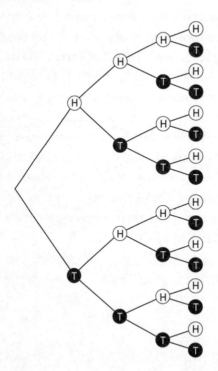

**Figure 1**

All possibilities for four consecutive coin tosses.

(each with probability 1/2). Whichever of these happens, the second toss is also either H or T (each with probability 1/2). Whichever of these happens, the third toss is also either H or T (each with probability 1/2). And whichever of these happens, the fourth toss is also either H or T (each with probability 1/2). So we get a 'tree' with 16 possible routes through it. According to probability theory, each route has probability $1/2 \times 1/2 \times 1/2 \times 1/2 = 1/16$. This is plausible, because there are 16 routes, and each should be equally likely.

Notice that TTTT has probability 1/16, and HTHH (say) *also* has probability 1/16. Although HTHH looks 'more random' than TTTT, they have the same probability. It is the *process* of tossing a coin that is random, but this does not imply that the results must always look irregular. Usually they do – but that happens because most sequences of H's and T's don't have much pattern, not because patterns are forbidden.

If you toss a coin four times, then on average you get exactly two heads. Does this mean that two heads and two tails is highly probable? No. From Figure 1, there are 16 different sequences of H's and T's, and exactly six of them contain two heads: HHTT, HTHT, HTTH, THHT, THTH, TTHH. So the probability of exactly two heads is $6/16 = 0.375$. This is *less* than the probability of *not* getting exactly two heads, which is 0.625. With longer sequences, this effect becomes even more extreme.

Calculations and experiments of this kind make it clear that there is *no* 'law of averages' – by which I mean that the future probabilities of independent events are *not* changed in any way by what happened in the past.

However, there is an interesting sense in which heads and tails *do* tend to balance out in the long run – despite what I've just told you. It depends on what the phrase 'balance out' means. If you mean the numbers eventually end up *equal*, then you're barking up the wrong tree. But if you mean that the *ratio* of the numbers eventually gets very close to one, you're absolutely right.

To see what I mean, imagine plotting the excess of the number of H's over the number of T's by drawing a graph of the *difference* at each toss. You can think of this as a curve that moves one step upwards for each H and one down for each T, so my sequence TTTTHTHHHHHHTT THTTTH produces the graph of Figure 2.

That establishes the principle, but the picture may still make you think that the numbers balance out pretty often. Figure 3 shows a random walk corresponding to 100,000 tosses of a fair coin, which I calculated on a computer. Here heads spends a startling amount of time in the lead. The walk starts at position 0 at time 0 and moves either by +1 ('heads') or by −1 ('tails') with equal probability at each subsequent stage. Notice that there seems to be a definite 'drift' towards positive values for times 40,000 onwards.

However, this drift is not a sign that there is something wrong with the computer's random number generator, so that the chance of heads is greater than that of tails. This sort of wildly unbalanced behaviour is entirely normal. In fact, much *worse* behaviour is entirely normal.

Why? It so happened that this particular walk reached position 300 (that is, heads led tails by 300 tosses) after about 20,000 tosses of the coin. Precisely because coins have no memory, from that stage on the 'average' excess of heads hovers around 300 − in fact, the walk spends more of the subsequent time below 300 than above, with tails

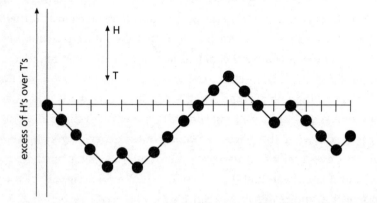

**Figure 2** Random walk representing excess of heads over tails.

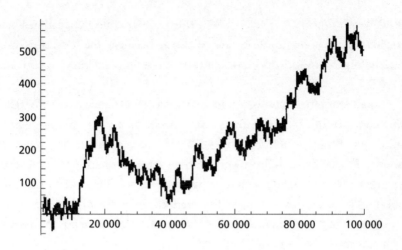

**Figure 3** Typical random walk for 100,000 tosses.

predominating from time 20,000 to about 80,000, but heads taking the lead again from toss 80,000 to toss 100,000.

Nevertheless, we can be sure that with probability 1 (virtual certainty) the walk will return to position 0 (equal numbers of heads and tails) *eventually*. However, because it has reached position 500 after about 100,000 steps, the time taken to get back to position 0 will probably be very long indeed. In fact, when this particular computer run was extended to 500,000 steps, the position ended up even further away from 0.

Notice the cluster of returns to 0 up to times 10,000. To be precise, this walk returned to 0 at times 3, 445, 525, 543, 547, 549, 553, 621, 623, 631, 633, 641, 685, 687, 1985, 1989, 1995, 2003, 2005, 2007, 2009, 2011, 2017, 2027, 2037, 2039, 2041, 2043, 2059, 2065, 2103, 3151, 3155, 3157, 3161, 3185, 3187, 3189, 3321, 3323, 3327, 3329, 3347, 3351, 3359, 3399, 3403, 3409, 3415, 3417, 3419, 3421, 3425, 4197, 4199, 4203, 5049, 5051, 5085, 5089, 6375, 6377, 6381, 6383, 6385, 6387, 6389, 6405, 6465, 6479, 6483, 6485, 6487, 6489, 6495, 6499, 6501, 6511, 6513, 6525, 6527, 6625, 6637, 6639, 6687, 7095, 7099, 7101, 7103, 7113, 7115, 7117, 7127, 8363, 8365, 8373, 8381, 8535, 9653, 9655, 9657, 9669, 9671, 9675, 9677, 9681,

9689, 9697, 9699, 9701, 9927, 9931, 9933 ... and no other times up to 500,000. (These numbers are all odd, because the positions are alternately even and odd, and start at the even number 0 when the time is 1.)

It may seem that, having got to an excess of 300 heads at time 20,000, the coin suddenly 'remembers' that it really needs to toss equal numbers of tails, so that by time 40,000 it has returned to an excess of about 30 heads. But why didn't it remember sooner? Or later? For example, at time 70,000, when again the excess of heads has risen to about 300, it looks as though the coin completely forgets that it is 'supposed' to toss equal numbers of heads and tails. Instead, the excess of heads moves relentlessly higher.

One 'pattern' is apparent: when it *does* get back to 0, we often see a cluster of such returns. For example, it returns at time 543, then 547, 549, 553. Or again, the return at 9653 is followed by 9655, 9657, 9669, 9671, 9675, 9677, 9681, 9689, 9697, 9699, 9701. This clustering happens because the walk is more likely to get to 0 quickly if it starts at 0. In fact, with probability 1/4 it will go from 0 to 0 in two steps.

Nonetheless, it will eventually escape to very distant parts of the number line indeed – as far from 0 as you wish, either in the positive or the negative direction. And having done so, it will eventually return to 0. But in these cases 'eventually' is entirely unpredictable, but typically very, very long.

Despite that, random walk theory also tells us that the probability that the balance *never* returns to zero (that is, that H stays in the lead *forever*) is 0. This is the sense in which the 'law of averages' is true – but it carries no implications about improving your chances of winning if you're betting on whether H or T turns up. Moreover, you don't know *how* long the long run is going to be – except that it is likely to be very long indeed. In fact, the average time taken to return to equal numbers of heads and tails is infinite! So the idea that the next few tosses will react to a current *excess* of heads by producing more tails is nonsense.

However, the *proportions* of heads and tails do tend to become closer and closer to 50%. Usually. Here's how. Suppose you toss a coin 100 times and at that stage you have 55 H's and 45 T's – an imbalance of 10 in favour of H's. Then random walk theory says that if you wait long enough, the balance will (with probability 1) correct itself. Isn't that the 'law of averages'? No. Not as that 'law' is normally interpreted. If you choose a length in advance – say a million tosses – then random walk theory says that those million tosses are unaffected by the imbalance. In fact, if you made huge numbers of experiments with one million extra tosses, then on average you would get 500,055 H's and 500,045 T's in the combined sequence of 1,000,100 throws. On average, imbalances *persist*. Notice, however, that the *frequency* of H changes from $55/100 = 0.55$ to $500055/1000100 = 0.500005$. The proportion of heads gets closer to $1/2$, as does that of tails, even though the difference between those numbers remains at 10. The 'law of averages' asserts itself not by removing imbalances, but by swamping them.

However, that's not quite the whole story, and as told so far it's unfair to people who claim that the numbers of heads and tails should eventually become equal.

According to random walk theory, if you wait long enough then eventually the numbers *do* balance out. If you stop at that instant, you may imagine that your intuition about a 'law of averages' is justified. But you're cheating: you stopped when you got the answer you wanted. Random walk theory also tells us that if you carry on tossing for long enough, you will reach a situation where the number of H's is a million more than the number of T's. If you stopped *there*, you'd have a very different intuition! A random walk just drifts from side to side. It doesn't remember where it's been, and wherever it's got to, it will eventually drift as far away from that as you wish. *Any* degree of imbalance will eventually happen – including none!

So it all depends on what we mean by 'eventually'. If we specify the number of tosses in advance, then there is no reason to expect the number of heads to equal the number of tails after the specified number of tosses. But if we can choose the number of tosses according to the result we obtain, and stop when we're happy, then the numbers of heads and tails 'eventually' become equal.

I mentioned earlier that the situation for dice is rather different. To see why, we need to generalize the random walk concept to more dimensions. The simplest random walk in the plane, for example, takes place on the vertices of an infinite square grid. A point starts at the origin, and successively moves one step either north, south, east, or west, with probability 1/4 for each. Figure 4 shows a typical path. A three-dimensional random walk, on a cubic grid in space, is very similar, but now there are six directions – north, south, east, west, up, down – each with probability 1/6.

It can again be shown that for a two-dimensional random walk, the probability that the path *eventually* returns to the origin is 1. The late Stanislaw Ulam (Los Alamos), best known for his co-invention of the hydrogen bomb, proved that in three dimensions the situation is different. Now the probability of eventually returning to the origin is about 0.35. So if you get lost in a desert and wander around at random, you'll eventually get to the oasis; but if you're lost in space and wander

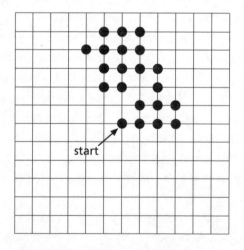

**Figure 4** Random walk in two dimensions.

around at random, there is only a chance of about one in three that you'll get back to your home-world.

We can use this random walk to tackle the die problem. Suppose we label the six directions of a three-dimensional random walk according to the faces of a die: north = 1, south = 2, east = 3, west = 4, up = 5, down = 6. Repeatedly roll the die, and move through the grid in a given direction whenever that face of the die turns up. In this experiment, 'return to the origin' means 'the same number of 1's as 2's, and the same number of 3's as 4's, and the same number of 5's as 6's.' The probability that this eventually happens is therefore 0.35. So the stronger condition 'all six numbers occur equally often' must have probability less than 0.35.

Even the simplest one-dimensional random walk has many other counter-intuitive features. Suppose you choose a large number of tosses in advance, say a million, and watch whether heads or tails is in the lead. What proportion of the time, on average, would you expect heads to take the lead? The natural guess is 1/2. Actually, this is the least likely proportion. The most likely proportions are the extremes: heads stays in front the whole time, or none of the time! For more information, see William Feller, *An Introduction to Probability Theory and Its Applications*.

## FEEDBACK

Feller's book states that in a two-dimensional random walk on a square grid, the probability of eventually returning to the origin is 1, but on a three-dimensional cubic grid the probability is less than 1, roughly 0.35. Several readers pointed out that the figure given in Feller's book is not quite correct. David Kilbridge of San Francisco told me that in 1939 the English mathematician George N. Watson gave the value as

$$\frac{1}{[3(18 + 12\sqrt{2} - 10\sqrt{3} - 7\sqrt{6}\,)(K(2\sqrt{3} + \sqrt{6} - 2\sqrt{2} - 3)]^2}$$

where $K(z)$ is $2/\pi$ times the complete elliptic integral of the first kind with modulus $z^2$.

If you don't know what that is, you probably don't want to! For the record, elliptic functions are a grand classical generalization of trigonometric functions like sine and cosine, which were very much in vogue a century ago and are still of interest in a number of contexts. However, they are seldom studied in today's undergraduate mathematics courses.

The numerical value is approximately 0.34053729551, closer to 0.34 than Feller's figure of 0.35.

Kilbridge also calculated the answer to my 'dice eventually even out' problem: they do so with probability approximately 0.022. For 'dice' with 2, 3, 4, 5 sides the analogous probabilities are 1, 1, 0.222, and 0.066.

Yuichi Tanaka, an editor for our Japanese translation, used a computer to work out the probability of eventually returning to the origin on a four-dimensional hypercubic grid. After running for three days, his program printed out the approximate value 0.193201673. Is there a formula like Watson's? Are there any elliptic function experts out there?

# Arithmetic and Old Lace

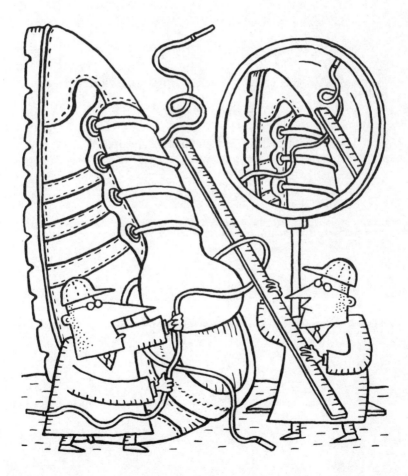

Which method of tying shoelaces uses the shortest amount of lace? A simple model leads to some remarkable geometry, and provides a definitive answer . . . except for various practical considerations, of course. Not only that: it's all done by mirrors.

**W**hat is mathematics? One proposal, made in desperation, is 'what mathematicians do'. By the same token a mathematician is 'someone who does mathematics', a neatly circular piece of logic that fails to pin either the subject or its practitioner down. Some years ago, in a rare flash of insight, it dawned on me that a mathematician is somebody who sees an *opportunity* for doing mathematics where others might not – just as a businessperson is someone who sees an opportunity for doing business where others might not.

To drive home the point, consider shoelaces. The potential for extracting significant mathematics from shoelaces is not widely recognized. That it exists was made clear to me by an article, 'The shoelace problem', written by John H. Halton of the Computer Science Department of the University of North Carolina in the *Mathematical Intelligencer*.

There are at least three common ways to lace shoes, shown in Figure 5: American zigzag lacing, the European straight lacing (from which the epithet 'strait-laced' derives, though perhaps by way of garments rather than shoes), and quick-action shoe-store lacing. From the point of view of the purchaser, styles of lacing can differ in their aesthetic appeal and in the time required to tie them. From the point of view of the shoe manu-facturer, a more pertinent question is which type of lacing requires the shortest – and therefore cheapest – laces. In this chapter I shall side with the shoe manufacturer, but readers might care to assign a plausible measure of complexity to the lacing patterns illustrated, and decide which is the simplest to tie.

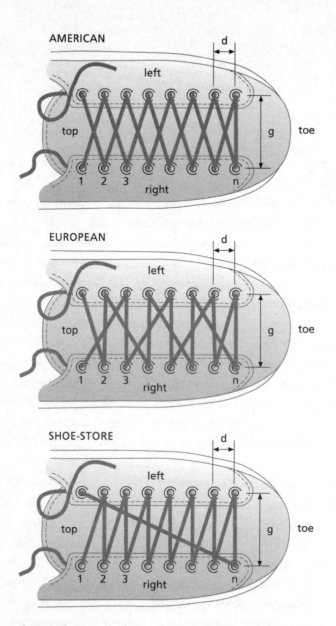

**Figure 5** Three common lacing patterns. The parameters $n$ (number of holes), $d$ (distance between successive holes), and $g$ (gap between pair of corresponding holes) are also shown.

Of course, the shoemaker is not restricted to the three lacing patterns shown, and we can ask a more difficult question: which pattern of lacing, among *all* the possibilities, requires the shortest lace? Halton's ingenious methods answer this too – subject to some assumptions, and the usual mathematical modelling simplifications like 'infinitely thin laces' – as I shall indicate nearer the end of this chapter.

I shall focus only on the length of shoelace that lies between the 'top' two eyelets of the shoe, on the left of the diagrams – the part represented by straight line segments. The amount of extra lace required is essentially that needed to tie an effective bow, and is the same for all methods of lacing, so it can be ignored. My terminology will refer to the lacing as seen by the wearer (hence my use of the word 'top' just now), so that the upper row of eyelets in the figure lies on the left side of the shoe, and the lower row on the right. I shall also idealize the problem so that the lace is a mathematical line of zero thickness and the eyelets are points. The big assumption made here is that the lacing is 'alternating'; that is, the lace alternates between left and right eyelets. It's still possible to do the sums if this assumption fails to hold, but to keep the analysis simple, we'll restrict attention to alternating lacings.

Using a brute force attack, the length of the lace can then be calculated in terms of three parameters of the problem:

- The number $n$ of pairs of eyelets.
- The distance $d$ between successive eyelets.
- The gap $g$ between corresponding left and right eyelets.

With the aid of Pythagoras's Theorem (one wonders what the great man would have made of this particular application) it is not too hard to show that the lengths for the lacings in Figure 5 are as follows:

*American:* $g + 2n \sqrt{(d^2 + g^2)}$
*European:* $ng + 2\sqrt{(d^2 + g^2)} + (n - 1)\sqrt{4(d^2 + g^2)}$
*shoe-store:* $ng + n\sqrt{(d^2 + g^2)} + \sqrt{(n^2d^2 + g^2)}$.

Which is the smallest? Suppose, for the sake of argument, that $n = 8$ as in the figure, $d = 1$, and $g = 2$. Then simple arithmetic shows that the lengths are:

*American*: $2 + 16\sqrt{5} = 37.777$
*European*: $16 + 2\sqrt{5} + 7\sqrt{8} = 40.271$
*shoe-store*: $16 + 8\sqrt{5} + \sqrt{68} = 42.134$.

In this case the shortest is American lacing, followed by European, and finally by shoe-store. But can we be certain that this is always the case, or does it depend upon the numbers $n$, $d$, and $g$?

Some careful high school algebra, using the formulas stated above, shows that if $d$ and $g$ are non-zero and $n$ is at least 3 then the shortest lacing is always American, followed by European, followed by shoe-store. If $n = 2$ and $d$ and $g$ are non-zero then American is still shortest but European and shoe-store lacings are of equal length. If $n = 1$, or $d = 0$, or $g = 0$ then all three lacings are equally long, but only a mathematician would worry about such cases!

However, this approach is complicated and offers little insight into what makes different lacings more or less efficient.

Instead of using complicated algebra, Halton observes that a clever geometrical trick makes it completely obvious that American lacing is the shortest of the three. With a little more work and a variation on that trick it also becomes clear that shoe-store lacing is the longest. Halton's idea owes its inspiration to optics, the paths traced by rays of light. Mathematicians discovered long ago that many features of the geometry of light rays can be made more transparent – if that is the word to use when discussing light – by applying carefully chosen reflections to straighten out a bent light-path, making comparisons simpler.

For example, to derive the classical law of reflection – 'angle of incidence equals angle of reflection' – at a mirror, consider a light ray whose path is composed of two straight segments: one that hits the mirror, and one that bounces off. If you reflect the second half of the path in the mirror (Figure 6) then the result is a path that passes through the front of the mirror and enters Alice's mirror-world behind the looking glass. According to the Principle of Least Time, a general property of light rays enunciated by Pierre de Fermat (yes, the poser of the 'Last Theorem'), such a path must reach its destination in the shortest time – which in this case implies that it is a straight line. Thus the 'mirror angle' marked in the figure is equal to the angle of incidence – but it is also obviously equal to the angle of reflection.

Figure 7 shows geometric representations of all three types of lacing, which Halton derives by an extension of this optical reflection trick. The figure requires a little explanation. It consists of $2n$ rows of eyelets, spaced distance $d$ apart in the horizontal direction. Successive rows are spaced distance $g$ apart vertically, and in order to reduce the size of the figure we have now reduced $g$ from 2 (as it was in Figure 5) to 0.5. The method works for any values of $d$ and $g$ so this causes no difficulty. The first row of the diagram represents the left-hand row of eyelets. The second row of the diagram represents the right-hand row of eyelets. After that, rows alternately represent the left-hand eyelets and the right-hand eyelets, so that the odd-numbered rows

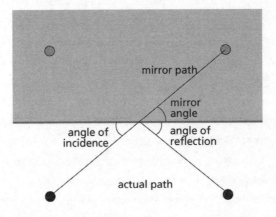

**Figure 6** By reflecting the path of a light ray in a mirror we can deduce the law of reflection from Fermat's Principle of Least Time.

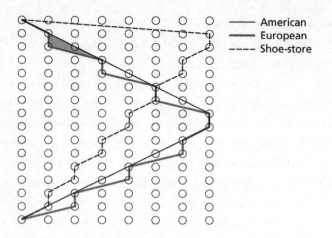

**Figure 7** Geometric representation of the three lacings obtained by successive reflections, here shown for the case $d = 1$, $g = 0.5$. By considering triangles like the one shown shaded it is evident that the American lacing is shorter than the European one.

represent left-hand eyelets and even-numbered rows represent right-hand eyelets.

The polygonal paths that zigzag across this diagram correspond to the lacings, but with an extra 'twist' – almost literally. Start at the top left eyelet of a lacing pattern and draw the first segment of lace, running from left to right of the shoe, in between the first two rows of the diagram. Draw the next segment of lace *reflected* to lie between rows 2 and 3 instead of going back from row 2 to row 1 as it does in a real shoe. Continue in this manner, reflecting the physical position of each successive segment whenever it encounters an eyelet. (Note that after two such reflections, the segment will be parallel to its original position but two rows lower, and so on.) In effect, the two rows of eyelets are replaced by mirrors. So, instead of zigzagging between the two rows of eyelets, the path now moves steadily down the figure, one row at a time, while its horizontal motion along the rows repeats precisely the motion along the rows of eyelets of the corresponding segments of the lacing pattern.

Because reflection of a segment does not alter its length, this representation leads to a path that has exactly the same length as the corresponding lacing pattern. The added advantage, however, is that it is now easy to compare the American and European patterns. In a few places they coincide, but everywhere else the American pattern runs along one edge of a thin triangle (one such triangle is shown shaded) while the European one runs along two edges of the same triangle. Because any two sides of a triangle exceed the third side in length (that is, a straight line is the shortest path between two given points), the American lacing is obviously shorter.

It is not quite so obvious that the shoe-store lacing is longer than the European. The simplest way to see this is to eliminate from both paths all vertical segments (which contribute the same amount to both lengths because each path has $n-1$ vertical segments) and also any sloping segments that match up. The result is shown in Figure 8 (thick lines). If each V-shaped path is now straightened out by reflection about a *vertical* axis placed at the tip of the V (fine lines) it finally becomes easy to see that the shoe-store path is longer, again because two sides of a triangle exceed the third side.

For the shoelace problem, this cunning combination of graphical representations and reflection tricks can do more than just compare particular lacing patterns. Halton uses it to demonstrate that the American zigzag lacing is the shortest among *all* possible lacing

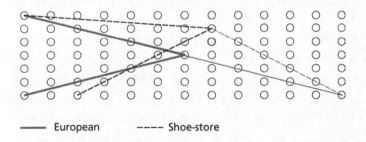

**Figure 8** Eliminating common segments of lace and reflecting in the vertical axis as well, to show that shoe-store lacing is longer than European.

patterns: the proof can be found in his article. More generally, both shoelaces and Fermat-style optics become united in the mathematical theory of geodesics – shortest paths in various geometries. There the reflection trick comes into its own in a big way, and Alice's mirror-world sheds light on fundamental questions in physics, as well as confirming the superiority of the American way of lacing shoes.

## FEEDBACK

Several readers queried the conclusion that the American way of lacing shoes uses least lace. This is true on the assumption that the lace passes alternately through the holes on the left and right side of the shoe. But if this assumption is removed, then shorter lacings can be obtained – though for practical reasons stronger string is needed. Frank C. Edwards III of Dallas, TX found two shorter methods when $n$ is even, both of length $(n-1)(g+2d)$: they are shown in Figure 9, with some segments bent for clarity. When $n = 18$, $d = 1$, $g = 2$ the length is 28, compared to 33.3 for the American pattern.

The second of these was also sent in by Michael Melliar-Smith of Santa Barbara, CA, Stephen Wallet of San Diego, CA, and several others. Neil Isenor of Waterloo, ON recalled being shown the same method by an officer cadet room-mate in the 1950s. William R. Read of Vancouver, BC told me that 'as a Canadian infantryman during the Second World War, I was required to lace my boots' in the same manner, adding that the method was known as 'Canadian straitlacing', and offering a similar method when $n$ is odd.

Maurice A. Rhodes of Nelson, BC punningly wrote that 'while I may be a heel to wax tongue in cheek and not be instep with our author, I would be straight-laced if I did not refute these claims. . . . Perhaps I can shoe-horn in another opinion.' He traced the method back to the Scots, asking whether I had mislaid my ancestry. (I should perhaps explain that despite my name, the earliest Scottish ancestor I can trace is a great-great-great-grandfather, a sea-captain by the name of Purves, who is

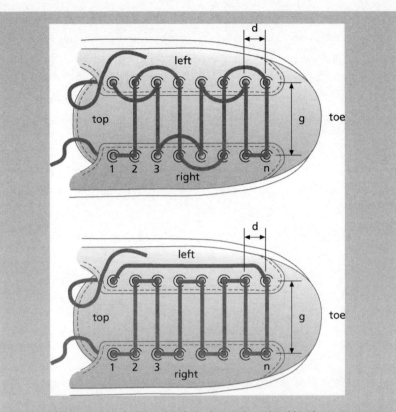

**Figure 9** How to beat American lacing by not alternating sides of the shoe.

buried in Canterbury Cathedral.) Rhodes explained that the same method was taught to air cadets at Canada's Royal Military Colleges in the late 1940s. In the Royal Canadian Navy, seamen lashed their boots in this manner, because 'a quick slash of the external lacing by a Seaman's knife . . . and the boots could easily be shucked off to avoid drowning. In the Royal Canadian Air Force and the Canadian Army the same method was used because a boot could be quickly and easily removed from an injured foot.' And Donald Graham of Vancouver, BC told me that his ten-year-old daughter Nicole invented the method for herself the first time she had to put new laces in her sneakers.

# 4

# Paradox Lost

Everything written on this page, up to the next full stop, is a lie. So it's not a lie, so it is a lie . . . Oops. The 'liar paradox' puzzled the ancient Greeks, and it still causes trouble today, with good reason. On the other hand, some other famous paradoxes don't really stand up to scrutiny.

$S$**ome of the most provocative** foundational problems in mathematics lie in the area of logic, which looks entirely straightforward but in actuality is littered with pitfalls. The great bugbear of mathematical logic is the existence of simple but baffling paradoxes. In everyday terms, a paradox may just be something that seems true and is actually false, or seems false and is actually true.

For instance 'the 21st century began in the year 2000' is widely thought to be true, but actually false. (The 1st century began in year 1, not year 0, because there *wasn't* a year 0. Now add 2000, and the 21st century began in the year 2001. Which is in fact correct, and why the movie *2001: A Space Odyssey* was not titled *2000: A Space Odyssey*.) Or the mathematical fact, known as the Banach–Tarski Paradox, that a solid sphere of unit radius can be cut into a finite number of disjoint pieces which can then be reassembled to make *two* solid spheres of unit radius. It seems obviously false because the volume should not change ... but the 'pieces' concerned are so complicated that they do not possess well defined volumes. But I digress.

From the mathematical point of view, these are relatively feeble paradoxes – they may force us to revise our views on some topic, but they don't force us to revise our *way* of thinking. The deepest logical paradoxes are self-contradictory statements. The simplest of these is the statement 'This sentence is a lie'. If the statement is true, then it tells us that it is false; if it is false, then it tells us that it is true. Worrying.

Paradoxes like this one forced mathematical logicians to be very careful indeed in defining what they were talking about, and what you were allowed to do with it. Bertrand Russell's 'barber paradox' is a case in point. In some village there is a barber who shaves everyone who does not shave themselves. Who shaves the barber? In the real world, there are plenty of get-outs, such as: are we talking about shaving beards here, or legs, or what? Is the barber a woman? Can such a barber actually exist anyway?

In mathematics, these easy ways out of the paradox are not available, and a more carefully stated version of Russell's paradox knocked the skids out from under the life's work of Gottlob Frege, who thought he had managed to base the whole of mathematics on the logical properties of sets. A *set* is a collection of objects, and the set is said to *contain* each such object. For instance, the set of all even numbers between 0 and 10 inclusive contains the objects 0, 2, 4, 6, 8, 10, and no others. Frege had assumed that any apparently sensible mathematical property defines a set, consisting of those objects that do have that property. But Russell invited Frege to contemplate a set (which we call $X$) defined as 'the set of all sets that do not contain themselves'. This is an apparently reasonable property. Some sets (for example, the set of all sets) do contain themselves. Others, such as the set of even numbers described earlier, do not contain themselves (the set concerned is not an even number between 0 and 10 – it's a *set*, not a number, OK?).

Very well, said Russell: does the set $X$ contain itself?

If $X$ contains $X$, then $X$ (in its role as a member of $X$) satisfies the defining property of not containing itself, so $X$ does not contain $X$.

On the other hand, if $X$ does not contain $X$, then $X$ (in its role as a set) satisfies the defining property of not containing itself, so $X$ does contain $X$.

Oops.

There are many paradoxes in the mathematical and logical literature. Some of these paradoxes stand up under scrutiny, and when they do, they illuminate the limitations of logical thinking (Paradox Regained). Others, including some that are traditional in recreational mathematics, don't fare so well (Paradox Lost). Or do they? Here are my views on a few of them, but you may disagree. If so, let's agree to differ: please don't write or e-mail to argue your case – life is too short.

My first paradox concerns the Greek lawyer Protagoras, who lived and taught in the 5th century BCE. He had a student, and it was agreed that the student would pay for his teaching after he had won his first case. But the student didn't get any clients, and eventually Protagoras threatened to sue him. Protagoras reckoned that he would win either way, for if the court upheld his case then the student would be required to pay up, but if Protagoras lost, then by their agreement the student would have to pay anyway. The student argued exactly the other way round: if Protagoras won, then by their agreement the student did *not* have to pay, but if Protagoras lost, the court would have ruled that the student did not have to pay.

All great fun, but I don't think this one stands up to scrutiny. Both litigants are doing a pick-and-mix – at one moment assuming the agreement is valid, then assuming that the court's decision can override it. *But why do you take an issue like this to court?* Because the court's job is to resolve any claimed ambiguities in the contract, override the contract if need be, and tell you what to do. So if the court orders the student to pay up, then he has to; and if the court says he doesn't have to pay up, then Protagoras doesn't have a leg to stand on. Legally, the court's decision takes precedence over the contract. Paradox Lost.

A far deeper paradox is due to Jules Richard, a French logician. It dates from 1905. Here's one version of it. In the English language, some sentences define positive integers and others do not. For example 'The year of the Declaration of Independence' defines the number 1776, whereas 'The historical significance of the Declaration of Independence' does not define a number. So what about this sentence: 'The smallest number that cannot be defined by a sentence in the English language containing fewer than twenty words.' Observe that whatever this number may be, we have just defined it using a sentence in the English language containing only nineteen words. Oops.

What's going on this time? The only obvious way out is if the proposed sentence does not actually define a number at all. However, it ought to. If we accept that the English language contains a finite number of words, then the number of sentences with fewer than twenty words is itself finite. For instance, if we allow 99,999 words, then there are at most $20^{100,000} - 1$ sentences of twenty words or fewer. (By allowing a blank word we increase the 99,999 to 100,000 but we can then include all shorter sentences in the total. The $-1$ removes the empty sentence – all blanks.) Of course many of these sentences make no sense, and many of those that *do* make sense don't actually define a positive integer – but that just means we have fewer sentences to consider. Between them, the sentences define a finite set of positive integers, and it is a standard theorem of mathematics that in such circumstances there is a unique smallest positive integer that is not in the set. So on the face of it, the sentence does define a positive integer.

But, of course, it can't.

Possible ambiguities of definition such as 'A number which when multiplied by zero gives zero' don't let us wriggle off the logical hook. If any sentence is ambiguous, then we rule it out: by 'define' we surely require an unambiguous result. Is the troublesome sentence ambiguous, then? Not really. The sentence is not troublesome because it fails to define a *unique* number. It is troublesome because it does not define a number at all. It looks as if it ought to – but the existence of such a number is logically contradictory, so the sentence can't actually define a number. Notice that if instead we had considered the very similar sentence 'The smallest number that cannot be defined by a sentence in the English language containing fewer than nineteen words' we wouldn't have had a problem. So the Richard paradox tells us something quite deep about the limitations of language as a description of arithmetic, namely: there is no easy way to determine, from the form of a linguistic statement, whether it is meaningful. Paradox Regained.

In a more recreational vein, there is the 'surprise test' paradox. Teacher tells the class that there will be a test one day next week

(Monday through Friday), and that it will be a surprise. This seems reasonable: the teacher can choose any day, and there is no way that the students can predict that day in advance. However, the students reason like this. The test can't be on Friday, because if it was, then as soon as Thursday passed without a test, we'd know it had to be Friday, so no surprise. But once we've ruled out Friday, we're down to the same set-up with a four-day week (Monday to Thursday), so by the same argument the test can't be on Thursday either. In which case it can't be on Wednesday, so it can't be on Tuesday, so it can't be on Monday – so no surprise test is possible.

On the other hand, if the teacher sets the test on Wednesday, there seems to be no way the students could actually know this ahead of time. So something is screwy about the logic. Is this a case of Paradox Lost or Paradox Regained? I think it's a very interesting case of something that looks like a paradox but isn't. There is a logically equivalent statement which is obviously true and totally uninteresting. Suppose that each morning the students announce confidently 'The test will be today.' Then eventually they will do so on the day of the actual test, at which point they will be able to claim the test was not a surprise.

I don't see any objection to this strategy on the part of the students, except that it's a cheat. The reason it works is that if *every* day you expect the surprise to happen, then of course you won't be surprised. My view – and I've argued with enough mathematicians who didn't agree, let alone anybody else, so I'm aware that there's room for differences of opinion – is that the alleged paradox of the surprise test is just this obvious strategy dressed up to look mysterious. It's not an obvious cheat because everything is intuited instead of being acted upon, but in reality it's the *same* cheat dressed up.

Let's sharpen the conditions by asking the students to say, each morning before school begins, whether they think the test will be that day. In order for the students to *know* that it can't be on Friday, they have to allow themselves the option of announcing on Friday morning: 'It'll be today.' And the same goes for Thursday, Wednesday, Tuesday, and Monday. So they need to say 'It'll be today' five times in all – once per day. Fair enough: if the students are allowed to revise their prediction each day, then eventually they'll be right.

If we demand a little more, though, the students' argument falls to bits, and so does the paradox. For example, suppose they are allowed only *one* such announcement. If Friday arrives and they haven't used up their guess, then they can indeed make the announcement then. But if they *have* used up their guess, they're in trouble. However, they *can't* wait until Friday to use their guess, because the test might be on Monday, Tuesday, Wednesday, or Thursday. In fact, if they are allowed *four* guesses, they're still in trouble. Only if they are permitted five guesses can they guarantee to predict the correct day.

If I showed you five boxes, one of which contained a large sum of money while the others were empty, and you had a rock-solid method for predicting the correct box with just one guess, I'd be impressed. But if your method required five guesses to succeed, I would be mightily unimpressed. You might make all five guesses up front, by pointing to each box in turn. You might use your guesses one at a time – point to box 1, open it, then point to box 2 if the first box was empty, and so on. Either way, I don't think anyone would be the least bit surprised when eventually you got the right box. Basically, I'm claiming that what the students are really doing is just a disguised version of this trivial kind of 'prediction'.

In fact, I'm suggesting two things here. The less interesting one is that the 'paradox' hinges on what we mean by 'surprise'. The more interesting claim is that *whatever* reasonable thing we mean by 'surprise', there are two logically equivalent ways to state the students' prediction strategy. One – the usual way to present the puzzle – seems to indicate a genuine paradox. The other – presented in terms of actual actions, not hypothetical ones – turns it into something correct but unsurprising, destroying the element of paradox entirely.

If you're not yet convinced, here's one final remark. Equivalently, we can up the ante by letting the teacher add another condition. Suppose that the students have poor memories, so that any work they do on a given evening to prepare for the test is forgotten by the next evening. If, as the students claim, the test is not going to be a surprise, then they ought to be able to get away with very little homework. Just wait until the evening before the test; then cram, pass, and forget. But the teacher, in her wisdom, knows that they can't do that. If they don't do their homework on Sunday evening, the test could be on Monday, and if so, they'll fail. Ditto Tuesday through Friday. So despite never being surprised by the test, the students have to do five evenings of homework.

Paradox Lost, I'd say.

## FEEDBACK

Several readers drew my attention to a fascinating article by David Borwein (University of Western Ontario), Jonathan Borwein, and Pierre Maréchal in the *American Mathematical Monthly*. These authors define a measure of surprise, and ask what strategy the teacher should follow in order to maximize surprise. They conclude that the day of the test should be chosen at random, where the probability of choosing a particular day of the week follows a precise pattern. (They allow 'weeks' of any integer length.) The probability remains roughly constant throughout the early part of the week, but increases rapidly on the last few days, with the final day having the highest probability. So even if you disagree with me and think the exam can't be a total surprise, it's possible to say how surprising it is.

Mind you, I haven't quite worked out how the degree of surprise changes if the students taking the test have read Borwein, Borwein, and Maréchal's article . . .

R. B. Burckel of Kansas State University sent a resolution of the Richard Paradox. Recall that this paradox is about the phrase 'The smallest number that cannot be defined by a sentence in the English language containing fewer than twenty words.' Whatever this number may be, the phrase defines it using a sentence in English containing only nineteen words. Yet it looks like such a number must exist: make a list (necessarily finite) of all possible phrases with nineteen or fewer words, strike out those that do not define a unique number, and take the smallest number that's omitted. However, there are problems with this argument: the list itself is not well defined, as Richard pointed out in a paper in *Acta Mathematica* in 1906. As a sample of the pitfalls, the list will include the following two phrases (I have modified Burckel's suggestions here and take responsibility for the result):

- The number named in the next expression, if a number is named there, and zero if not.
- One plus the number named in the preceding expression.

Each expression *on its own* seems to define a number unambiguously, so should be kept in the list. But the two together are contradictory. Notice that the order of the phrases in the list matters here – and this problem is just the tip of a nasty self-referential iceberg. Since the list is not well defined, the paradoxical phrase does *not* define a unique number, so this becomes a case of Paradox Lost. And it's not possible to recover the paradox by insisting that self-referential networks of phrases are all considered to be ambiguous, and hence are removed from the list. Changing the list also changes what networks are self-referential, so there is no consistent way to get an unambiguous list.

# Tight Tins for Round Sardines

Obviously, you can pack 49 milk bottles, of unit diameter, into a square crate whose side measures seven units. Just use seven rows of seven. But can you pack the same bottles into a *smaller* square crate, by arranging them differently? Want to bet?

The party game 'sardines' involves packing as many people as possible into a closet, and it's aptly named. Mathematicians like to play sardines, too, but both people and fish are too awkward a shape to think about, so they prefer circles. What, they ask, is the smallest size of square crate into which you can pack 49 milk bottles? Or, equivalently, given a unit square, what is the diameter of the largest circle such that 49 copies of that circle can pack inside the square without overlapping? To see that these questions are equivalent, by the way, note that in the first case we fix the size of the circles and vary that of the square, while in the second case it's the other way round. So, apart from the choice of scale, solving one problem automatically solves the other. Provided, of course, that I don't put milk bottles in upside down or on their sides, and assuming that the cross section of a milk bottle can be modelled as a perfect circle, and the cross-section of a crate as a perfect square.

Questions like this ought to be as old as mathematics itself, but nearly all of the information that we now have about them dates from 1960 or later. The reason is that 'combinatorial geometry', as this area is known, is surprisingly subtle. Answers are far from obvious, and proofs are hard to come by. For example, it is surely obvious that the smallest square crate into which you can pack 49 milk bottles of unit diameter has a side of seven units: just arrange the bottles in a square array (see Figure 10a).

Obvious this may be, but – like many allegedly obvious ideas – it is false. In 1997 K. J. Nurmela and P. R. J. Östergård found a way to pack 49 circles inside a slightly smaller square (Figure 10b). The difference in

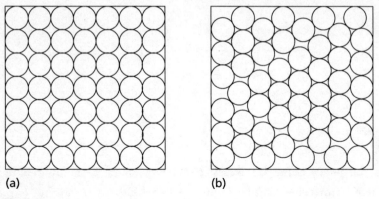

(a)                                          (b)

**Figure 10**

(a) The obvious way to pack 49 circles inside a square.

(b) If instead they are packed like this, the square becomes (very slightly) smaller.

sizes is so small that it is invisible to the naked eye. This packing disproved a conjecture of G. Wengerodt, who had already shown that the obvious square packing is optimal for 1, 4, 9, 16, 25, and 36 circles, but not for 64, 81, or any larger square number. Wengerodt left the case of 49 circles open, but guessed – wrongly, it now turns out – that the square packing was still densest.

You may be wondering why the square packing is not the densest one for *any* square crate, however large. From the right point of view, it is easy to see that the square packing must fail if the crate is large enough. You have to know (this is easy to check) that in the infinite plane there is a packing that is more dense than the square lattice, namely the hexagonal packing – like the arrangement of balls at the start of a game of pool, but infinitely extended.

A finite-sized crate has a square boundary, and this prevents the formation of a perfect hexagonal lattice, which is why square arrangements are densest for some small numbers of circles. But when the number of circles gets big enough, the effect of the boundary becomes so small that solutions close to a hexagonal lattice can pack in more circles than the square arrangement. This is how Wengerodt proved that the square packing is not the best for square numbers greater than or equal to 64. The case of 49 circles is rather delicate, though, which is why it took a while to find the right answer.

I was prompted to write about such matters when I received a copy of *Packing and Covering with Circles* by Hans Melissen, a doctoral thesis examined at the University of Utrecht in December 1997. It is by far the best and most complete survey of such questions that I know, and it contains many new arrangements and proofs, and a comprehensive list of references. Similar questions can be asked for regions of many different shapes – circles, rectangles, triangles. There are many potential applications, from industrial packaging to the physics of electrons. However, the real charm of the topic lies in its elegant mathematics.

The problem of packing equal circles inside a square, maximizing the size of the circles relative to the side of the square, seems not to have been discussed in print before 1960, when Leo Moser conjectured a solution for eight circles. His conjecture was verified soon afterwards, and it led to a series of publications on the same question with different numbers of circles. In 1965 J. Schaer, one of the mathematicians who proved Moser's conjecture, published solutions for up to nine circles. He remarked that optimal packings for up to five circles are easy, and attributed the solution for six circles to Ron Graham (now at Bell Labs).

It is usual to reformulate the problem slightly, so that the circles themselves disappear from consideration. If two equal circles touch, then their centres are separated by a distance equal to their common diameter. And if a circle touches a straight boundary, then its centre lies on a line that is parallel to the boundary, but is separated from it by a distance equal to the radius of the circle. So my question about circles can be rephrased as: 'Place 49 points in a given square, in such a way as to maximize the minimum separation between any two of them.' The points here are what used to be the centres of the circles; the square is not the original square, but a smaller one, whose sides have been moved inwards by an amount equal to the radius of a circle. The advantage of the 'point' formulation is its conceptual simplicity. In that formulation, the current state of play for up to 20 circles is summarized in Figure 11. All arrangements shown there have been proved optimal. For 17 points there are two distinct arrangements. Some arrangements, such as those for 13, 17, and 19 points, involve 'free' points whose position is not completely fixed, but can vary within certain (small) limits.

A trickier variant concerns packing circles (or, again, equivalently points) inside a circle. The earliest known publication on this question is the 1963 Ph.D. thesis of B. L. J. Braaksma, on a technical question in analysis. In among the technicalities, he conjectures an optimal arrangement for eight points. (It is curious that in both problems the case of

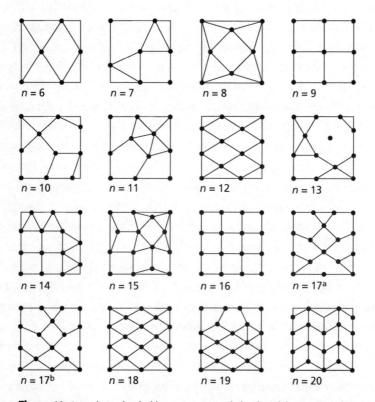

**Figure 11** Arranging points inside a square to maximize the minimum separation.

eight points seems to have been the first to attract serious attention.) Later, he found a proof that his arrangement is the best, but never published it. In this formulation the solution is known for 11 or fewer points. Optimal arrangements for between 12 and 20 points have been conjectured, but proofs are lacking. (See Figure 12). Again, there are alternative arrangements in several cases (6, 11, 13, 18, and 20). For six points the two solutions are (a) five points on the boundary, with some freedom to move, plus one at the centre, and (b) a perfect hexagon. The conjectured solution for 19 points is especially elegant and symmetric.

The proof for 11 points was first provided by Melissen, in the afore-mentioned thesis. His method is to begin by partitioning the circle into a system of curiously shaped regions, and using estimates of distances to show that some of these regions contain at most one of the points that are to be distributed inside the circle. In this way, the investigator gradually gains 'control' over the disposition of the points – establishing in this case, for example, that eight of the points must lie on the boundary of the circle. The method is delicate, and relies on an intelligent choice of the partition; however, it is reasonably general and some version of it can be used on many such problems, often with the aid of extensive computer calculations.

Packings inside an equilateral triangle are especially interesting, because this shape of boundary relates rather neatly to the hexagonal lattice – as any pool player knows. The wooden or plastic triangle used to set up the balls initially is an equilateral triangle, and the balls pack together inside it as part of the hexagonal lattice. In fact, such packings were first studied only when the number of circles (or equivalently points, as always) was a triangular number: 1, 3, 6, 10, 15, and so on. Such numbers are of the form $1 + 2 + 3 + \ldots + n$, and in these cases the circles can be arranged in part of a perfect lattice packing. The hexagonal lattice is known to be the optimal arrangement over the whole plane, a fact widely assumed but first proved by Axel Thue in 1892. It is therefore highly plausible that the optimal packing of a triangular number of points inside an equilateral triangle is the obvious

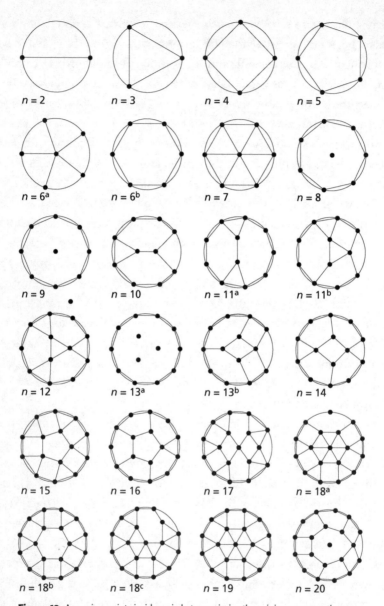

**Figure 12** Arranging points inside a circle to maximize the minimum separation.

pool-ball arrangement. This is in fact true, but a proof is quite tricky: Melissen gives a particularly neat one. He also finds (and proves) optimal arrangements for 12 points or fewer, together with conjectures for 16, 17, 18, 19, and 20 points (Figure 13).

The whole area has a pristine beauty that is very appealing, but it also shows how deceptively simple problems of this type can be. This is not

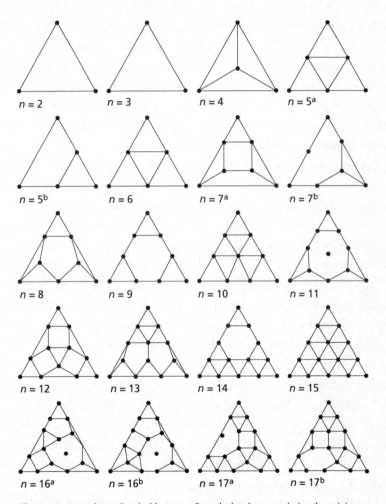

**Figure 13** Arranging points inside an equilateral triangle to maximize the minimum separation.

an easy area for a 'serious' mathematician to work in. In fact it is better suited to the recreational mathematician, for whom there are innumerable attractive challenges: prove some of the conjectures, improve upon them and thereby disprove them, extend conjectured or proved solutions to larger numbers of points . . . The shape of the domain can be changed, too: there are some known results for rectangles and for isosceles right triangles, for instance. Hexagons look like fun.

The packing question can even be posed on curved surfaces. In 1930 the Dutch botanist P. M. L. Tammes asked for optimal packings of circles on the surface of a sphere. Melissen considers a variant of the Tammes problem, using not a sphere but a hemisphere (Figure 14). Here the results are proved for 6 or fewer points, but only conjectured for 7 to 15 points. For those feeling really ambitious, what about packing spheres in three-dimensional regions?

I mentioned potential applications to physics. In 1985 A. A. Berezin published a short note in *Nature* about minimum-energy configurations of identical electrically charged particles inside a disc. This has the same mathematical flavour as circle-packing, because the particles will repel each other, which is much like trying to maximize their minimum separation. However, this analogy should not be taken too literally, because what really counts here is energetic balance, not separation *as such*. What the system actually does is minimize its total energy. At any rate, the prevailing intuition was that all the charges should repel each other until they reached the edge of the disc, a conclusion generally justified by a result known as Reverend Earnshaw's Theorem. This states that no charged body can be in equilibrium under electrostatic forces only, so that equilibrium requires the imposition of conditions at the body's boundary. Berezin's numerical calculations, however, showed that for between 12 and 400 electrostatic charges, the distribution with one at the centre and the rest at the boundary has lower energy than that with them all at the boundary.

The discrepancy between physical intuition and Berezin's computations was eventually resolved in favour of physics – though there was

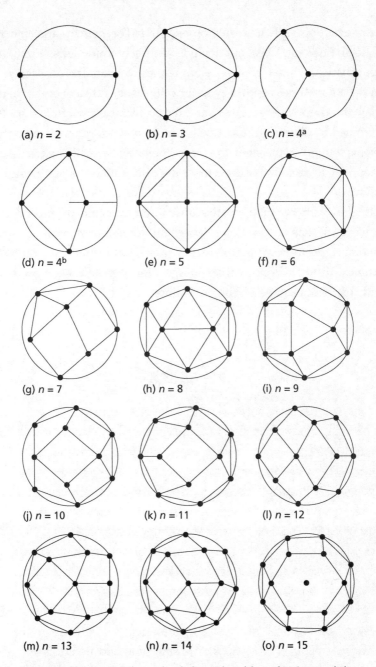

**Figure 14** Arranging points on a hemisphere (viewed from above) to maximize the minimum separation.

nothing wrong with Berezin's observation. The point is that the real physical universe does not include infinitely thin discs. Either the mathematical model represents a two-dimensional cross-section of parallel line charges inside a cylinder, or the disc will have finite, though small, thickness. In the former case, the correct energy differs from that computed by Berezin (it must be based on a logarithmic force law, not inverse square). In the latter, the central point will actually migrate a tiny amount away from the true centre of the disc, until it reaches the nearby boundary!

In this manner, mathematics and intuition were reconciled. The problem still has much interest, however; for example, Melissen gives the first rigorous proof that Berezin's numerical results are correct. So, technical difficulties notwithstanding, a lot of progress is now being made on these elegant, baffling questions.

## FEEDBACK

It was made clear to me that I ought to have clarified my remark that 'nearly all the information we have about such questions dates from 1960 or later', when one reader complained – with surprising bitterness – that I had cavalierly disregarded the great classical work of Gauss, Lagrange, and others, and thereby denigrated the intellectual heritage of Western thought. By 'such questions' I meant packing objects into *finite* regions, like a square crate. The classical work is on packings of the infinite plane, and it also assumes that the objects form a regular array. The essence of the problems discussed in the chapter is that the region is of limited extent, and there are no regularity assumptions about how the circles are placed.

Several mathematicians and physicists sent me their research papers. One, by Kari Nurmela (Helsinki University of Technology), discusses a related but subtly different problem, mentioned near the end of the chapter: distribute point charges in a circular disc so as to minimize the total energy (with inverse square law repulsion). The paper is referenced in Further Reading. It lists the best known configurations for all numbers of point charges up to and including 80 (previously only numbers up to 23, plus 29, 30, and 50, had been considered). As might be anticipated on physical grounds, the points space themselves approximately in a series of concentric rings.

# 6

# The Never-Ending
# Chess Game

The laws of chess include some obscure rules that are intended to prevent pointless games that go on forever. The idea of a tripleless sequence, which first arose in dynamics, shows that one reasonable proposal to change these laws is not up to the task. In fact, it would let you play forever without moving any pawns.

**A**nyone who plays chess knows that some games just peter out into pointless situations where neither player looks like winning, nothing constructive can be done, and there's no obvious way to end the game except to agree a draw. But what if the other player won't agree? Then the game might go on indefinitely. The bodies who frame the laws of chess have foreseen such situations, and many different rules have been proposed to force games to end. The classic law is 'the game shall be drawn if the Player prove that 50 moves have been made on each side without checkmate having been given and without any men having been captured or Pawn moved'.

However, some recent computer analyses have shown that there are some end-games where one player can force a win, but this involves making more than 50 moves without capturing any pieces or moving pawns, so the laws of chess are forced to specify certain exceptional situations. Any law that specifies a limit on the number of moves permitted under particular conditions runs exactly the same risk, so it would be nice to find a different approach altogether.

One proposal, made some time ago, was that the game should end if the same sequence of moves, in exactly the same positions, is repeated three times in a row. (Do not confuse this with the standard law that if the same *position* occurs three times, the player facing it can claim a draw. But note that this law does not oblige them to do so.) It might be a short sequence, or a long one: the proposed rule was careful not to specify the length.

You can make out a good case that any violation of this three-in-a-row rule ought to end the game. The question is: are there pointless games that do *not* violate it? It's here that the mathematical worldview sees an interesting question. Can a game of chess go on forever, without checkmate, and without repeating the same sequence of moves three times in a row? (A game that goes on forever is certainly a pointless one.)

Chess is kind of complicated, so any mathematician worth their salt would try to simplify. Suppose we focus on just two possible moves, represented by the binary symbols 0 and 1. Can a sequence of 0's and 1's go on forever without any finite block repeating three times in a row?

It turns out that there are many ways to produce such a sequence, which I'll call a *tripleless sequence*. The first was invented by Marston Morse and Gustav Hedlund while investigating a problem in dynamics. Begin with a single 0. Follow it by the complementary sequence (every 0 changed to a 1 and vice versa), which here is just 1, so you get 01. Then follow that by its complementary sequence, and so on, building up an infinite sequence like this:

0
01
01**10**
0110**1001**
01101001**10010110**

and continuing the process indefinitely. I've written the complementary sequences in boldface for clarity.

This sequence is genuinely tripleless, but a proof of that property is tricky. There is a more explicit tripleless sequence for which the proof is a bit easier. In order to describe it, we need some terminology. Recall that an even number is a multiple of 2, whereas an odd number is one greater than a multiple of 2; more simply, even numbers are of the form

$2m$ and odd numbers of the form $2m + 1$. We need a similar terminology for multiples of *three*. Say that a number is

> *treble* if it is a multiple of 3 (that is, of the form $3m$);
> *soprano* if it is one higher than a multiple of 3 (that is, of the form
>   $3m + 1$);
> *bass* if it is one lower than a multiple of 3 (that is, of the form
>   $3m - 1$).

Every whole number is either treble, soprano, or bass. If a number is
soprano (equal to $3m + 1$ for some $m$) then say that $m$ is its *precursor*.
For example $16 = 3 \times 5 + 1$ is soprano, and its precursor is 5, which is
bass.

Using this terminology, we can write down a recipe for a sequence
that never repeats a block three times in a row:

- The first term is 0.
- The $n$th term in the sequence is 0 if $n$ is treble.
- The $n$th term in the sequence is 1 if $n$ is bass.
- If $n$ is soprano, with precursor $m$, then the $n$th term in the
  sequence is equal to the $m$th term.

The first three rules tell us that the sequence goes

> 010*10*10*10*10 . . .

where the pattern *10 repeats indefinitely, and the starred entries
aren't yet determined. The fourth rule lets us work upwards along the
starred entries. For example, entry 4 is the same as its precursor, which
is entry 1, and that's a zero. Entry 7 is the same as its precursor, which
is entry 2, and that's a one; and so on. Because the precursors are
smaller, their values will have been worked out already, so the fourth
rule does indeed determine all the stars.

These rules lead to what I'll call the *choral sequence*:

010 **010** 110 **010** 010 110 **010** 110 110 **010** 010 110 . . .

I've grouped the terms in threes to show the structure more clearly, and put the soprano terms in boldface. The choral sequence has the curious property that the boldface terms reproduce the entire sequence exactly.

There are lots of *double* repetitions of blocks in the choral sequence: for example, it starts 010 010, and the first 18 terms repeat the sequence 010010110 twice. But no block ever repeats *three* times (see the box), so it is tripleless.

How does this help with the chess problem? There are many more moves in chess than just two; and if you pick two (say advancing the king's pawn and moving the king's rook three spaces forwards) it's not at all clear that the sequence corresponds to *legal* moves. The way to get round this is actually quite simple; but you might like to think about it before reading on.

OK, here goes. Suppose that both players confine themselves to moving one or other of their knights out and back as in Figure 15. Depending on their current position, either the outward move or the backward move is available for each knight. Suppose the players use the sequence of 0's and 1's to determine their moves, with '0' interpreted as 'move the king's knight' (KN) and '1' as 'move the queen's knight' (QN), like this:

0 White moves KN (out)
1 Black moves QN (out)
0 White moves KN (back)
0 Black moves KN (out)
1 White moves QN (out)
0 Black moves KN (back)

and so on.

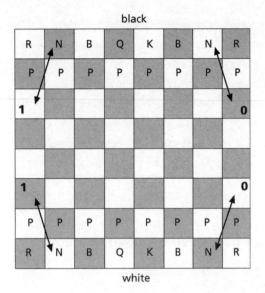

black

white

**Figure 15**

In the never-ending chess game, only knights ever move, back and forth between two squares. Symbols 0 and 1 show the corresponding terms in the tripleless 'choral sequence'.

It's not exactly an exciting chess game, but it's a perfectly legal one – in the sense that each individual move is legal. And because of its relation to the choral sequence, it is clear that it goes on forever without ever repeating the same sequence of moves three times in a row. In fact, more strongly, it doesn't repeat the same sequence of *pieces* (KN or QN) three times in a row. So if you're looking for a truly watertight chess law to terminate pointless games – one that is proof even against players colluding to play stupidly but legally – that old proposal doesn't work.

This particular problem motivates mathematicians to ask related questions about symbol sequences. For example, is there a sequence of 0's and 1's that never repeats a block *twice* in a row? Does the answer change if you're allowed more symbols, say 0, 1, 2? Recreational mathematicians can have fun turning such questions into analogous ones about chess; for example, can a legal game of chess continue indefinitely without any block of moves repeating twice in a row?

The effect of all this mathematics on the framing of chess laws is unlikely to be great, however, because chess players normally have a sensible objective in mind and we don't (yet?) know how to capture that condition mathematically.

To define something is to draw very tight boundaries around it. My own view is that everything really interesting has fuzzy edges, which just get fuzzier when you try to pin them down with a formal definition. Indeed, it is this fuzziness of boundaries that provides lawyers with a living: even such apparently black-and-white concepts as 'dead' or 'female' turn out to have fuzzy boundaries. Despite this, every chess player knows what a 'sensible' game of chess is – even though they can't *define* 'sensible'.

●●●●●●●●●●●●●●●●●●●●●●●●●●●●●●●●●●●●●●●●●●●●●●●●●●●●●●●●●●●●●●●●●●●

### PROOF THAT NO BLOCK OCCURS THREE TIMES IN A ROW

Call successive symbols 0 or 1 the terms of the sequence, and say that the $n$th term is treble, bass, or soprano if $n$ is.

1. No block of length 1 is repeated three times, because any three consecutive terms must include both a treble term and a bass term, which are different.
2. No block of length 2 is repeated three times, because any six consecutive terms contain a block of the form 0*1, but neither 010101 nor 101010 does.
3. If a block of length 3 is repeated three times, then it contains three soprano terms whose precursor terms are all the same and consecutive – which is ruled out by step 1.
4. If a block whose length is a multiple of 3 – say $3k$ – is repeated three times, then a similar argument shows that a block of length $k$ must have been repeated three times earlier in the sequence.
5. The only remaining case is when a block whose length is at least 4, and not a multiple of 3, is repeated three times. In this case the proof gets more complicated. To see the idea, suppose that the length is 4, so the sequence includes a block of the form *abcdabcdabcd*. One of the first three terms must be treble; suppose for example that it is *c*. Then the block actually goes *ab***0***dab***0***dab***0***d*. But every third term after the first 0 – marked in bold – is also treble, so $b = a = d = 0$ and the entire block goes 000000000, which is ruled out by step 1. Similar arguments hold if *a* or *b* is treble. A more convoluted version of the same kind of argument works for any block whose length is not a multiple of 3.

●●●●●●●●●●●●●●●●●●●●●●●●●●●●●●●●●●●●●●●●●●●●●●●●●●●●●●●●●●●●●●●●●●●

## FEEDBACK

The saga of the never-ending chess game prompted a number of letters clarifying the history of the 'Morse–Hedlund' tripleless sequence 0110100110010110 . . .

Jeffrey Shallit at the University of Waterloo wrote (I have edited his letter slightly, and left the references in place rather than removing them to Further Reading): 'Nowadays this sequence is usually attributed to the Norwegian mathematician Axel Thue, who wrote about the repetition problem in a series of papers beginning 1906. He also proved it was *overlap*-free, a stronger property. The application to chess was, to my knowledge, first observed in Morse's abstract in *Bulletin of the American Mathematical Society* **44** (1938) 632. For a humorous view of Morse's application see D. McMurray, "A mathematician gives an hour to chess" in *Chess Review*, October 1938. It was reprinted in Bruce Pandolfini (ed.) *The Best of Chess Life & Review* 1 (1933–1960), p. 84. Recently several mathematicians observed that the sequence was implicitly contained in an older paper by E. Prouhet in *Comptes Rendus* **33** (1851) 225.'

I. J. Good at Virginia Tech noted that Machgielis ('Max') Euwe, World Chess Champion 1935–7, invented the same sequence in 'Set theory observations on chess', *Proceedings of the Academy of Sciences of Amsterdam* **32** (1929) 633–42. He adds 'This article provoked me to invent (in 1943 or 44) the "reflection order" for the teleprinter five-unit code: for details see my article "Enigma and Fish" in *Codebreakers* (eds. F. H. Hinsley and Alan Stripp), Oxford University Press 1993. This code is now called the Gray code and was independently invented and patented by F. Gray for analog-to-digital conversion.'

# 7

# Quods and Quazars

Can you form a square before your opponent does? You each have 20 pieces of your own special colour, plus six white ones. And the board has 117 squares. Oh, and don't forget: once you decide to play one of your quods, you can't use any more of your quazars.

# G.

**G. Keith Still is a computer scientist** whose main professional interest is in the simulation of crowd dynamics and the design of appropriate control barriers. Keith is a very inventive person, and some years ago he came up with a mathematical game, which he calls Quod.

Quod is played on an $11 \times 11$ grid of cells from which the four corners have been removed, which leaves 117 cells available. The two players, black and red, each have 20 pieces of their own colour (called quods) and six white pieces (quazars). The players take turns to place one of their quods on the board in an unoccupied cell. The aim of the game is to get four of your quods at the corners of a square. Such a square may have its edges parallel to those of the board, or it may be tilted (Figure 16). To claim a win, the player shouts 'quod!' Usually this happens when the fourth and final corner of the square is placed in position, but players sometimes overlook a square that has formed accidentally, in which case they may shout 'quod!' as soon as it is their turn. However, this cannot be done after the other player has shouted 'quod!' – oversights must be rectified while the game is still in progress.

Figure 17 shows how black can place a whole series of quods so that at each stage red is forced to make a unique move in order to prevent the formation of a winning square. This kind of forced play does not make for a very interesting game, which is where the other type of piece, the quazar, comes in. Quazars are purely for blocking and do not count towards the formation of squares – which is why they are white for both players. The rules for playing quazars are:

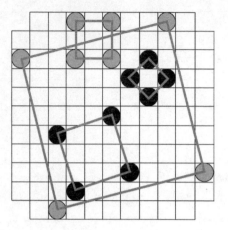

**Figure 16**

A few of the thousands of possible winning squares.

1. It must be your turn to move.
2. You may play as many quazars as you wish (up to your limit of six) but they must all be played before playing your quod.
3. You must then play a quod as usual, after which your turn ends.

Finally, there are two technical rules. If on the last move of the game a player leaves a position that would have forced a win on the next move (had there been any quods left to play) then that player is declared the winner anyway. And if the game ends without either

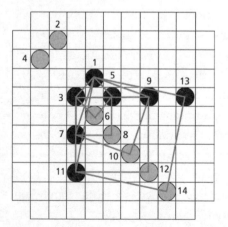

**Figure 17**

A sequence of moves that (in the absence of quazars) keeps red on the defensive.

player forming a square, then the player with the most quazars unplayed wins. (If both players have the same number of quazars unplayed, the game is drawn.)

Because a huge number of possible squares can be formed, Quod is a surprisingly complex game. It is, for example, relatively easy to make a 'double square' move – one that creates two potential squares simultaneously. By 'potential square' I mean a square with three of its four corners filled in and the last one currently unoccupied. Your opponent can block the completion of several potential squares in one move by using quazars, but it is good strategy to create double squares in order to force your opponent to use up their quazars. Experience shows that the central cell is a reasonable opening move. After that, you need to be on the lookout for potential (or accidentally completed!) squares in unusual orientations, and watch for overlapping squares that might lead to double square moves for you or your opponent. Figure 18 takes you though an illustrative game; various potential squares are shown as they arise.

There are many variants on Quod, which lead to enjoyable games in different circumstances.

*Reduced Board*. Young children will find the game more manageable on a smaller board, in which case the number of quazars should be reduced accordingly (that is, five quazars on a $10 \times 10$ board, four on a $9 \times 9$, and so on).

*More than Two Players*. The rules are similar, but the number of quazars is reduced. With three players, each has four quazars; with four players, each has three; and with five or six players, each has two.

*Quod Trek*. This is like the standard two-player game but each has only six quods and six quazars. After all quods are played, a turn consists of picking up a quod of your colour and replacing it elsewhere. Quazars may be played at any time as in the standard game, but once played they cannot be moved.

*Quod Duel*. Each player has quods of two (or more) colours and plays one quod of each colour per turn. They may shout

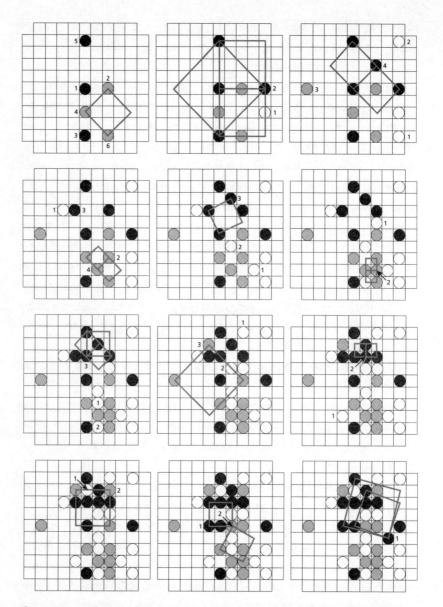

**Figure 18** A sample game of Quod. Any potential squares are drawn. Numbers indicate the order in which moves are played, starting from the position in the previous frame of the diagram. Here black wins by forcing a double square when red has run out of quazars.

'quod!' only when they have formed a square from four quods of the same colour.

*Quod Rapid.* Each player has six quods and six quazars. At each move they may either place a new quod or move an old one. Quazars are played as usual and cannot be moved once placed.

*Quod Bridge.* This requires four players, who form two pairs. Players sit opposite their partners round the four sides of the board. One pair of players have black quods, the other pair red. Play proceeds clockwise round the board. No conferring is allowed – you have to work out what your partner's strategy is before your opponents do. But you are permitted to send signals (waving arms, banging your head on the table, leaping up and down, and so on).

## FEEDBACK

I had mistakenly referred to the game as 'Quad', thinking of words like 'quadratic' associated with squares, but its inventor Keith Still reminded me that he favours the spelling 'Quod' as in 'Quod erat demonstrandum'. (I've corrected this error here.) The game acquired quite a following, and a disk version was given away in vast quantities with a computer magazine. David Weiblen programmed Quod himself, and set the computer playing hundreds of games, employing a strategy based upon weighting positions according to a set of rules that reflects their apparent strength.

In his simulations the first player always won. This leads him to question how interesting the game really is; it leads me to ask whether his weighting rules actually lead to the best play. He also pointed out that there are exactly 1173 possible squares, a figure confirmed by Les Reid of South West Missouri State University. Solutions were posted by Michael Kennedy (University of Kansas), Ken Duisenberg (Hewlett-Packard), and Denis Borris. Borris generalized the result to the $n \times n$ case, the answer being $(n^4 - n^2 - 48n + 84)/12$, and Duisenberg did the $m \times n$ case.

# 8

# Zero Knowledge Protocols

PINs, passwords, electronic signatures . . . these days it seems as if you can't even buy a newspaper without offering proof of your identity. And when you do, someone can observe your PIN, steal your password, or forge your signature. Here's an alternative: how to prove that you know something without revealing what it is.

**I**n the age of the Internet it has become important to be able to send messages that convey certain facts to their intended recipient without inadvertently revealing other facts, to them or to anybody else. For instance, suppose that you want to pay for a purchase by credit card. Transmitting your credit card number on its own is not a wise method. In order for this bare message to work, the recipient must transfer money whenever a valid credit card number is received. But then somebody could intercept your number, or even set up an illicit computer program that collects credit card numbers, and buy other goods using your account.

Using a simple PIN doesn't add much security, because that too must be transmitted over the net. Most security systems use some kind of encryption method to confirm that the message is from a legitimate source. Such systems work if the code is a secure one, and nowadays there are plenty of good ideas for secure codes. In fact, some codes are so secure that law-enforcement agencies want them banned, because they would let criminals send messages that could not be understood even if they were intercepted. On the other hand, civil liberties groups want the privacy of individuals to be protected against government snoops.

An alternative approach to encryption is to use a 'zero knowledge protocol'. This is a way of convincing the recipient that you are in possession of some key item of information (such as a PIN) without revealing what that item actually is. The surprise is that such protocols can exist at all, but in recent years cryptographers have devised them in large numbers.

The kind of principle that is involved can be illustrated most simply in the context of map-colouring rather than PINs. The celebrated Four Colour Theorem was first conjectured in 1852 by Francis Guthrie, a graduate student at University College London. It states that every map in the plane can be coloured so that no two adjacent countries have the same colour, and no more than four colours are used. It was eventually proved in 1976 by Kenneth Appel and Wolfgang Haken of the University of Illinois. However, if we are limited to only three colours, then some maps can be coloured and some cannot.

Suppose that your bank manager sends you an exceedingly complex map, and you wish to convince her that you know how to three-colour it *without* revealing which regions have which colours. Then you can construct an elaborate electronic device, linked to and controlled by two touch-sensitive screens, one in the bank, one at your home. This device is configured to do the following, and *only* the following, things. (See Figure 19.)

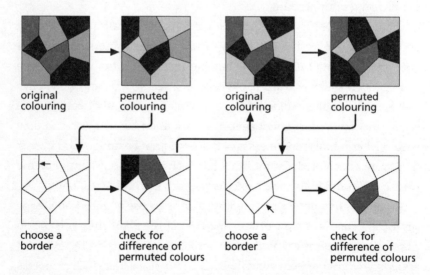

original colouring   permuted colouring   original colouring   permuted colouring

choose a border   check for difference of permuted colours   choose a border   check for difference of permuted colours

**Figure 19** Convincing your bank manager that you can three-colour a map. Repeat until all borders have been selected.

First, you program into the machine your map-colouring (say by touching regions of your screen – one touch for red, two for blue, three for yellow).

Next, the bank manager selects a border where two countries meet. The machine performs a random permutation of your colouring scheme – for example, systematically replacing your red colour by blue, your blue by red, and leaving yellow unchanged. There are six possible ways to permute the colours, and your bank manager does not know which permutation the machine has selected. Then the manager's screen displays the *new* colours of the two countries adjacent to the selected border, all other countries being left uncoloured. If your original colouring was a valid one, then these two colours should be different.

The manager then repeats the same operations until every border has been tested, and can then determine whether your claim to have three-coloured the map is correct. In fact, if your original colouring is faulty, with two adjacent countries having the same colour, then at some stage the bank manager will select their common border and the two permuted colours revealed by the machine will be identical. If, on the other hand, the two permuted colours are different for every border, then your original map must be a valid one.

However, because the permutations are random, there is no way for the manager to deduce your original colouring. The machine's responses merely confirm that various pairs of adjacent countries have different colours on your map: they don't tell her what the colours are.

Workers in the field of zero knowledge protocols prefer a more rigorous argument based on the idea of 'simulation'. Imagine a superficially identical set-up, in which the machine's responses are determined not by your choice of map, but by choosing two different colours at random and putting them on the screen. This fake system might produce many different sequences of pairs of colours, but one of the possibilities is the actual sequence of responses based on your map. Suppose for a moment that your bank manager could determine your map from the real machine's responses. Then she could also determine your map on that rare occasion when the fake machine produced the same responses. But for the fake machine, there is no such thing as 'your map', so such a deduction must be impossible.

Observe now that if your bank manager cannot deduce your three-colouring from the machine's responses, then neither can an illegitimate observer. Observe also that the bank manager has to believe that the machine really does work the way I've described, and isn't just displaying random pairs of colours.

A more elaborate zero knowledge protocol allows you to convince your bank manager that you know the two prime factors $p$ and $q$ of a particular number $n = pq$, without revealing what they are. Provided $n$ is fairly large – a typical size is around 200 digits – then there is no known algorithm that will find the factors $p$ and $q$ within the lifetime of the universe. However, there are very quick algorithms to test $p$ and $q$ to ensure that they are prime.

So your bank manager can cook up two large primes $p$ and $q$, work out $n = pq$, and treat $p$ and $q$ as a kind of PIN (which you are told when you open your account). Over a suitable communication channel you can then convince her that you know this PIN, without divulging $p$ and $q$ to her or any eavesdropper. The method involves a certain amount of number theory (see the box) and requires a further technique known as 'oblivious transfer'.

An oblivious transfer channel lets you send your bank manager two encrypted items of information, in such a manner that (a) she can decipher and read exactly one of the items, (b) you don't know which item she can read, and (c) you are both convinced that (a) and (b) are true. Subject to a few plausible conjectures, there are simple number-theoretic ways to construct an oblivious transfer channel, but I won't describe them here. For more details see Neal Koblitz, *A Course in Number Theory and Cryptography*.

This method does require a certain amount of preparation, and your usual four-digit PIN is replaced by two 100-digit numbers, upon which you must carry out quite a lot of arithmetic, flawlessly. However, any laptop is more than capable of such a task. There are more practical methods than the one I've just outlined, but they aren't as simple to describe. What is clear is that in an age of digital communications, security systems must be *provably* secure: experiment alone is not sufficiently convincing. And once you start asking for proofs, you're talking mathematics.

## PROVING KNOWLEDGE OF PRIME FACTORS BY OBLIVIOUS TRANSFER

You both know a number $n$, which is a product $n = pq$ of two primes $p$ and $q$, and you both know $p$ and $q$. A trusted, independent source supplies you both with a sequence of random bits 0 or 1 from which you can construct any random numbers required in the protocol. You can convince your bank manager that you know $p$ and $q$, without revealing what they are. The method uses 'modular arithmetic' in which multiples of a particular number $n$, called the modulus, are identified with zero. Specifically, the notation $y \bmod n$ means the remainder on dividing $y$ by $n$, for integers $y$, $n$. With this notation, the method runs as follows.

1. The independent source generates a random integer $x$, and sends you and your manager the remainder $r$ on dividing $x^2$ by $n$ (that is, $r = x^2 \bmod n$).
2. According to number theory, $r$ has exactly *four* different square roots modulo $n$. You use your knowledge of $p$ and $q$ to find them. One of them is $x$, and the other three are $n - x$, $y$, and $n - y$, for some $y$. (If you don't know $p$ and $q$, there is no efficient algorithm to find these square roots; indeed, if you know all four then you can easily deduce $p$ and $q$.)
3. You choose one of these four numbers at random: call it $z$.
4. You choose a random integer $k$ and send your manager the integer $s = k^2 \bmod n$. You then set $a = k \bmod n$, $b = kz \bmod n$, and send these two numbers to your bank manager by oblivious transfer.
5. The manager can read exactly one of the two messages. She checks that its square mod $n$ is either $s$ (if she reads message $a$) or $rs$ (if she reads message $b$).
6. These steps are repeated $T$ times. At the end of this, your manager is convinced (with probability $1 - 2^{-T}$) that you know the factorization.

Notice that there is no communication from your bank manager back to you; that is, the protocol is not interactive.

# FEEDBACK

Tom Sales of Somerset, NJ sent me a comment inspired by zero knowledge protocols. Many years ago Martin Gardner introduced a card game called 'Eleusis' in which one player invented rules and the others had to guess them by being told whether a given play was legal or illegal. At that time Sales invented a similar game, 'Alpha', involving a mouse who inhabits a triangular room. In each of its three corners is an array of coloured light bulbs. Alpha is frightened by the lights, and scurries from corner to corner according to rules like 'If the light in my corner is red and the light clockwise from me is green then I will run to the clockwise corner'. One player sets up the rules, secretly, and the other(s) try to deduce them by setting combinations of lights and watching where the mouse moves. A crucial feature of the game is that the rules depend only on the state of the lights relative to the mouse's current position, so permuting the corners does not change the rules.

Now eliminate the mouse! If you can't see the mouse then there is no way to deduce the rules. But at any random instant the mouse can be rendered visible, so that an observer can check that the rules are indeed being followed. Mouse movements thus form the basis of a zero knowledge protocol. Now let the mouse's movements represent a message, so that the rules for moving the mouse act as an enciphering algorithm, and you have a very interesting system, with a zero knowledge flavour, for transmitting code messages. With suitable additional features – Sales recommends incorporating his code system 'Omega' – it appears virtually unbreakable.

# 9

# Empires on the Moon

In the distant future, every nation on Earth also owns a plot of land on the Moon. Naturally, the national leaders want the maps to depict both of their territories with the same colour. And to avoid confusion, adjacent territories, on Earth or Moon, should have different colours. What's the smallest number of colours that the mapmakers can use? Oddly enough, no one knows.

**M**athematics **intrigues people** for at least three different reasons: because it is fun (the most important reason for inclusion in this book), because it is beautiful, or because it is useful. There are degrees of utility: a mathematical idea or method may be useful elsewhere in mathematics, it may be useful to theoretical scientists, it may be useful on the laboratory bench, it may be useful out in the world of industry and commerce, or it may be useful to ordinary citizens in their everyday lives.

I don't think that a mathematical concept has to be *directly* useful to justify its existence, or even the expenditure of taxpayers' money: mathematics is a coherent, interlocking whole and advances in one area often lead to advances elsewhere – and *those* may be useful even if the original advance wasn't. But I always take especial pleasure when a mathematical idea which at first sight seems totally useless turns out to have direct practical utility. Such examples are the best arguments against trying to judge mathematics in terms of superficial appearances. They are among the reasons why such exercises as the 'golden fleece awards' for useless science are often superficial, foolish, and misguided.

The concept of an '*m*-pire' is a case in point. It looks like harmless (and rather pointless) fun, but it has serious uses – as we'll see in the next chapter. In this one I'll just introduce the idea and explain some of the mathematics.

Imagine that planet Earth has been carved up into separate nations, each owning one connected region of territory – land and sea.

Moreover, each Earthly nation has annexed a connected region of the Moon, to create an empire that consists of two connected regions: one on Earth, the other on its satellite. Between them, these regions cover both worlds completely. What is the smallest number of colours that will colour a map of *any* such disposition of territory, in such a manner that both countries in any particular empire receive the same colour, but no two adjacent regions receive the same colour – on either the Moon or the Earth?

The answer is unknown: it is either 9, 10, 11, or 12. It's a fun problem, but a highly artificial one.

A typically useless product of ivory tower intellectuals?

Not at all.

In 1993 Joan P. Hutchinson of Macalester College, St Paul, Minnesota published a thorough survey of such questions in *Mathematics Magazine*. In one section of the article she described an application of Earth/Moon colouring to the testing of printed circuit boards, discovered by researchers at AT&T Bell Laboratories, Murray Hill. The connection is not at all obvious, but it is easy to understand, and it involves some concepts that will interest recreational mathematicians and that, in any case, deserve to be more widely known. The main one is the so-called 'thickness' of a graph.

In this chapter I'll describe the mathematics of maps, empires, and graphs, and explain what 'thickness' is. In the next, we'll take a look at the application to electronic circuit boards.

A *map* is an arrangement of regions, either in the plane or on a surface such as a sphere. Each region is a single connected portion of the plane or surface, and the regions make contact along common boundaries, which are curves. Often we make additional assumptions – for example, that no region completely contains another region.

A *graph* is a diagram formed from a number of blobs, called *nodes* or *vertices*, which are joined together by a number of lines, known as *edges*. Graphs are simpler and more abstract than maps.

However, any map can be represented by assigning a node to each region and joining two such nodes by an edge if and only if the corresponding regions share a common stretch of border (Figure 20). Imagine the nodes as capital cities, and the edges as highways that join cities in adjacent countries, crossing at their common border. This is the *map graph*. It represents which regions share a common boundary with others, but removes from consideration various distracting complications, such as the shapes of the regions. For many questions, the shapes don't matter, and it's often easier to get rid of them altogether – hence the map graph.

A graph is said to be *planar* if it can be drawn in the plane without any edges crossing. If we start with a map in the plane, then its map graph is obviously planar. More surprisingly, if a map is drawn on the sphere, or on several disconnected planes and spheres – as is the case for Earth/Moon maps – then the resulting graph is *still* always planar. To see why, imagine a map drawn on a sphere. Put a node in each region, and whenever two regions have a common boundary, connect the corresponding nodes with edges. The result is a graph that can be drawn on a *sphere* without any edges crossing. However, any such graph can be opened up and spread out on a plane. To do this, imagine cutting a small hole in the sphere, which does not meet any of the nodes or edges of the graph. Now imagine that the sphere is made from elastic sheeting. You can pull that tiny hole, making it bigger and bigger. The rest of the sphere stretches and deforms, carrying the graph

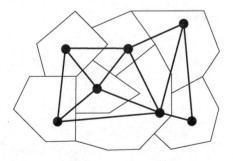

**Figure 20**

A map and the corresponding graph.

with it. By pulling it far enough you can flatten it out into a disk. Lay the disk on the plane, and you've now drawn the map graph on a plane without any edges crossing.

If the map is drawn on several spheres, we just do the same for each of them, and lay all the resulting disks out in the same plane without overlaps. The resulting graph will be disconnected – it will fall into several separate pieces, one for each sphere – but that's quite a common feature of graphs, and is allowed by their definition, so it doesn't matter.

An important graph for this chapter is the *complete graph $K_n$*, which has $n$ nodes, and an edge joining every pair of distinct nodes. Figure 21 shows $K_5$. If $n$ is 5 or larger, then the graph $K_n$ is *not* planar.

A map (on a plane, sphere, several spheres, whatever) is said to be *k-colourable* if its regions can be coloured, using no more than $k$ colours, so that regions that share a common boundary curve receive different colours. (Regions that meet only at a point, or finitely many points, can if necessary receive the same colour.) The analogous property for a graph runs along very similar lines. A graph is $k$-colourable if its nodes can be coloured, using no more than $k$ colours, so that nodes joined by an edge receive different colours. It is easy to see that a map is $k$-colourable if and only if its map graph is $k$-colourable. Just colour each capital city – each node of the graph – with the colour of the corresponding country.

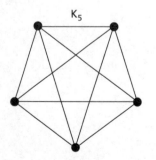

$K_5$

**Figure 21**

The complete graph $K_5$, which is not planar.

The smallest such $k$ is called the *chromatic number* of the graph: it tells us the minimum number of different colours needed for that graph – hence also for the corresponding map, if it is a map graph. Evidently $K_n$ has chromatic number $n$, because each node is joined to every other node, so no two nodes can be coloured the same.

Colouring problems have been the object of mathematical study for about a century. The best known result is the famous Four Colour Theorem, which says that every map in the plane can be 4-coloured. Percy Heawood proved long ago that every planar map can be 5-coloured: as remarked earlier the number was reduced to four in 1976 by Kenneth Appel and Wolfgang Haken in a tour de force that combined mathematical analysis with extensive computer searches and calculations. To this day, no proof that avoids heavy use of computers is known, although the Appel–Haken proof has been simplified considerably. Many generalizations have been studied too, among them the Earth/Moon maps that I mentioned at the start of this chapter.

A problem closely related to Earth/Moon maps was introduced by Percy Heawood in 1890. The problem is set on the Earth only, but now each country is part of an empire containing a maximum of $m$ countries, and the same colour must be used for every country in a given empire, again with adjacent regions having different colours. (Countries in a given empire are assumed not to touch each other.) Such a map is punningly known as an *m-pire*. Heawood proved that an $m$-pire can always be coloured with $6m$ colours, for all $m \geq 2$.

Since an $m$-pire is a particular type of map, it has an associated map graph with one node per country. However, it is no longer true that every legal colouring of the map graph corresponds to a colouring of the empire. The reason is that the standard colouring rules for a graph fail to fulfil the requirement that nodes from the same *empire* receive the same colour. It is difficult to handle this condition using the map graph. Instead, the construction of the graph is modified so that the colouring rules are automatically correct.

Here's how.

The *m-pire graph* associated with a given *m*-pire map has one node for each empire (not one for each region). If you find this confusing, think of the node as representing the emperor. Two nodes are joined by an edge if and only if the corresponding empires include at least one pair of adjacent countries. You might think of the *m*-pire graph as the 'invasion graph' of emperors whose empires can go to war across a common border. One node per emperor, one edge for every possible two-sided war.

Conceptually, the *m*-pire graph is obtained from the ordinary graph by identifying all the nodes in a given empire – drawing them in exactly the same place. This construction often leads to multiple edges – two nodes joined by several edges instead of just one. Superfluous edges of this kind are removed, to leave just one edge.

Identifying all the nodes in a given empire automatically forces them to receive the same colour, so the number of colours needed for an *m*-pire is the same as the chromatic number of its *m*-pire graph.

In 1983 Brad Jackson (San Jose State University) and Gerhard Ringel (University of California, Santa Cruz) used this approach to prove that the number 6*m* in Heawood's theorem cannot be reduced. They did this by demonstrating that you can find an *m*-pire whose *m*-pire graph is the complete graph $K_{6m}$. Since $K_{6m}$ definitely needs 6*m* colours, there is an *m*-pire that cannot be coloured with fewer than 6*m* colours.

There are connections between Earth/Moon maps and *m*-pire maps. In fact, an Earth/Moon map can be viewed as a particular kind of 2-pire, with a slightly curious underlying geometry (two spheres) that splits all the 2-pires into two pieces. Its graph consists of two disjoint planar graphs – for example, one possible arrangement is shown in Figure 22a. (The rounded shape has nothing to do with the Earth or Moon: recall that any graph on a sphere, or several spheres, can be deformed so that it lies in a plane. It's just easier to show the shape of the graph here using curved edges.)

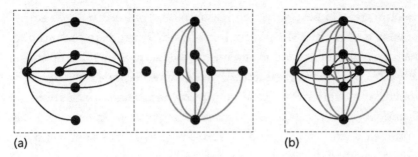

**Figure 22**

(a) Graphs for the terrestrial and lunar territories of a set of eight empires.

(b) Identifying corresponding nodes to create the corresponding 2-pire graph.

Suppose that we now think of this Earth/Moon graph as a 2-pire graph, so that nodes belonging to the same empire are identified to create Figure 22b. We see that the resulting graph need no longer be planar. Indeed, this one isn't.

However, the graph is 'almost planar'. The way it is constructed shows that its edges can be separated into two subsets, each of which forms a planar graph on the original set of nodes. Here the two subsets are the edges in Figure 22a and those in Figure 22b.

Such a graph is said to have thickness 2. In general, a graph has thickness $t$ if its edges can be separated into $t$ subsets, and no fewer, in such a manner that each subset forms a planar graph. Now, every map graph is planar, even when the map lives on a sphere. An Earth/Moon map is made up from two separate planar maps: one on the Moon, the other on the Earth. Each empire is represented exactly once in either of these maps. So every Earth/Moon graph has thickness 2: one planar bit for the Earth part, the other for the Moon part. The converse is also true: every graph of thickness 2 corresponds to an Earth/Moon map (although the territories involved may not completely cover the two worlds: there may be regions unclaimed by any of the empires).

Because an Earth/Moon graph is a special kind of 2-pire graph, Heawood's theorem implies that 12 colours are *sufficient* for any Earth/Moon graph. However, we can't conclude directly that 12 colours are also necessary. The reason is that not every 2-pire corresponds to an Earth/Moon map. In an Earth/Moon map, each empire has one region on the Moon and one on the Earth. If we think of this as a 2-pire, then the regions form two separate 'islands', and there is exactly one region from each empire on each island. In contrast, a 2-pire consists of a number of pairs of regions, which need not be arranged to form two islands – and even if they are, some empires might have both territories on the same island.

In fact, *none* of the known 2-pire graphs that actually require 12 colours can be turned into Earth/Moon maps. It therefore remains possible that *fewer* than 12 colours might always be enough for an Earth/Moon graph.

For instance, the complete graphs $K_9$, $K_{10}$, $K_{11}$, and $K_{12}$ are all 2-pire graphs, but they have thickness 3, and so cannot be Earth/Moon graphs (because those have thickness 2). In fact, the thickness of $K_n$ is 3 if $n = 9$ or 10, and is the greatest integer not exceeding (or 'floor' of) $(n + 7)/6$ otherwise.

Figure 22b is in fact the complete graph $K_8$, so $K_8$ has thickness 2. This means that it can be represented as an Earth/Moon graph. This proves that at least 8 colours are needed in the Earth/Moon problem. Rolf Sulanke (Humboldt University Berlin) increased this lower limit to 9 by showing that the graph of Figure 23 has thickness 2 and chromatic number 9.

The concept of thickness, then, is the deep mathematical idea that underlies the recreational puzzle of Earth/Moon maps. You might like to think about Earth/Moon/Mars maps, where every emperor has

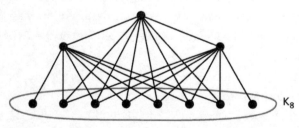

$K_8$

**Figure 23**

Sulanke's graph of thickness 2, which requires nine colours.

*three* territories, one on each world. These maps are particular kinds of 3-pire map, and their 3-pire graph always has thickness 3. In general a graph of thickness $t$ can be thought of as the $t$-pire graph of a system of galactic empires on a collection of $t$ planets.

Map-colouring problems of this kind are great fun – but they have little obvious practical significance. Even if we had planetary empires, the geographers could always colour their maps by trial and error – and in any case they might not want to follow our colouring rules. We shall see in the next chapter that there *are* applications of the concept of thickness; however, they are not literal translations of the 'map' image. Instead, they apply to the testing of electronic circuits.

Mathematics is abstract and general: the same idea has many realizations. Some are more fun than others – and some are more practical than others.

# 10

# Empires and Electronics

If you thought that the previous chapter was abstract nonsense, with no conceivable applications, think again. It leads to a remarkably efficient method for testing printed circuit boards to find short circuits. While the obvious method needs hundreds of thousands of tests, the one based on lunar empires takes less than a dozen.

T he **previous chapter took** a look at various map-colouring problems, relating them to graphs: diagrams in which dots known as 'nodes' are joined by lines known as 'edges'. A good mathematical idea has many different real-world interpretations. Although map-colouring problems appear frivolous, the mathematics behind them is useful to industry and commerce. In particular, the concept of the 'thickness' of a graph, to which we were led by the unlikely scenario of maps of empires on the Earth and the Moon, has recently been turned to advantage in the manufacture of electronic circuit boards. Now I will describe this application, which was discovered by researchers at AT&T Bell Laboratories, Murray Hill. It concerns testing boards to locate short circuits, and it is spectacularly efficient, reducing an impractically large number of tests, say 125,000, to a mere four.

Recall that a graph is planar if it can be drawn in the plane without any edges crossing each other. The next step up from planar graphs is graphs of thickness 2, meaning that their edges can be separated into two sets, in such a manner that either set, together with all the nodes, is planar. A graph has thickness 3 if its edges can be separated into three such sets, and so on. You can think of a graph of thickness 2 as a kind of 'sandwich'. On one slice of bread we draw the edges in the first set, none of them crossing; on the second slice, we draw the rest of the edges, again with none crossing. The nodes form the filling (Figure 24). A graph that needs $t$ layers of bread has thickness $t$.

This image makes it clear why graphs and their thicknesses are relevant to electronic circuits. To begin with, think of an electronic

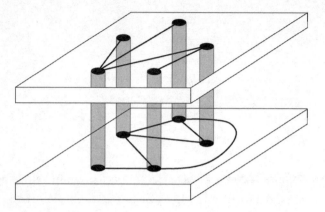

**Figure 24** The complete graph $K_5$ on five nodes represented as a sandwich. Each layer of bread is a planar graph, and the nodes form the filling. The graph $K_5$ appears if the sandwich is viewed from directly above, superimposing the two slices.

circuit as a graph in its own right. The nodes are the electronic components, and the edges are electrical connections. If the circuit is to be constructed on one side of a printed circuit board (PCB) then it must be planar to avoid short circuits. By using two sides of the board – like the two slices of bread in the sandwich – graphs of thickness 2 become available. By using several boards, the thickness of the graph can be increased. Similar considerations apply in the more hi-tech world of silicon chips, too, because VLSI (Very Large Scale Integrated) circuits have to be built in layers.

A typical PCB is a $100 \times 100$ array of holes – the exact numbers vary – where components can be attached, joined by horizontal and vertical lines that can be plated with 'tracks' of a conducting material, connecting the components together. An important problem for manufacturers of PCBs is to detect boards with spurious connections – extra bits of track that result in components being joined together electrically when they should actually be isolated from each other.

For practical reasons, manufacturers arrange the components on a PCB into 'nets'. A net is a collection of components, connected by tracks, so that the tracks contain no closed loops (Figure 25). In a well made PCB, distinct nets should not be electrically connected. The problem that concerns us here is to determine, in an efficient manner, whether two distinct nets have inadvertently been linked together by an unwanted bit of track – a 'short circuit'.

The most obvious way to do this is to check all pairs of nets to see whether they are connected. The simplest method here is to make a 'test device' that creates a circuit that runs from one net to the positive pole of a battery, and from the negative pole through a light bulb to the second net (Figure 26). If the two nets are inadvertently connected by the PCB's tracks, then current will flow and the bulb will light. If not, it won't. Of course, a practical test device would use more sophisticated electronics – such as a computer attached to a robot that automatically discards a faulty board, instead of a light bulb – but that's the basic idea.

Unfortunately, this approach is not practical. With $n$ nets, this method requires $n(n - 1)/2$ tests – the number of pairs of nets. Since 500 nets is typical, that means 125,000 tests per board, which is much

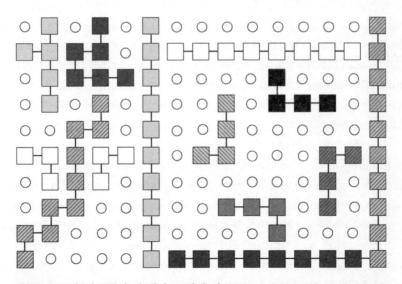

**Figure 25** A simple PCB circuit. Circles are holes for components, squares are components. Linked sets of squares are nets.

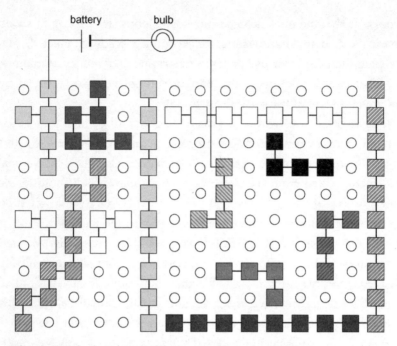

**Figure 26** Testing for a short circuit between a red net and a green net.

too big to be feasible. I will now convince you that applying the concept of the thickness of a graph quickly reduces the number of tests to a mere 11. In fact, a little extra thought reduces that number to just four. This means that every board that is manufactured can be tested quickly and efficiently, so that those with unintended short circuits can be discarded.

The starting point for these improvements is to turn the PCB design into a graph. The idea is to define the *simplest* graph that conveys information about short circuits between different nets: let me call this the *net graph* of the circuit design. The criterion of simplicity makes the construction of the net graph a little bit subtle. For example, because we are trying to find out whether or not there exist short circuits between different *nets*, there is no point in taking the nodes of the net graph to be the individual circuit components. Instead, we assign one node to each net. The edges of the net graph represent potential short circuits, not actual ones – because if we knew where the actual short circuits were, we wouldn't need to test the circuit. To be precise, two nodes of the net graph will be joined by an edge whenever the corresponding nets are 'adjacent' – meaning that they can be connected by a horizontal or vertical straight line that passes through no intermediate net (Figure 27).

Let me explain this choice, which is partly pragmatic.

In principle a short circuit might exist that connects non-adjacent nets. However, nearly all such short circuits must *also* connect adjacent nets, because of the way the circuits are built. In a typical manufacturing process, the fabrication device makes two passes over the board: one to create the horizontal connections, the other to create the vertical ones. Errors arise when it lays down too much conducting material, inadvertently linking two nets that should remain disconnected: I'll call such an error a 'fabrication fault'. There are other possible ways to create a short circuit and produce a faulty board, but they are far rarer than fabrication faults, and we can ignore them.

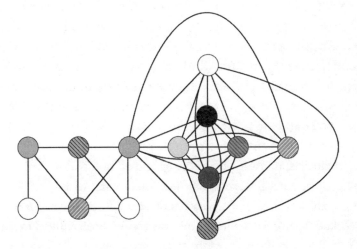

**Figure 27** The net graph for the circuit of Figure 25. Colours of blobs correspond to colours of nets. The graph has been 8-coloured so that no two adjacent nodes have the same colour. Heawood's theorem guarantees a similar colouring for any net graph, but perhaps needing up to 12 colours.

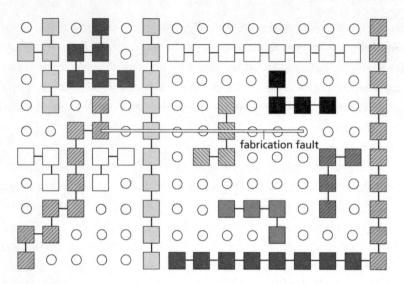

**Figure 28** Any short circuit caused by a fabrication fault must connect adjacent nets, even if it also connects others.

Because the connections are laid down as horizontal or vertical lines, any fabrication fault must create an unwanted link between two adjacent nets. The extra line of conducting material may run across several further nets, but the first two that it links will necessarily be adjacent (Figure 28). In other words, we can detect fabrication faults by looking for short circuits between *adjacent* nets. In this sense, the edges of the net graph correspond to the possible mistakes in fabrication. The condition about there being no intermediate nets simplifies the graph, but does not lose sight of any possible mistake: instead of looking for all short circuits, it just looks for the 'minimal' ones.

I said earlier that the graph whose nodes consist of the PCB's *components* has thickness 2 – one for each side of the PCB. The net graph also has thickness 2, for the same reason. I also mentioned a theorem proved by Percy Heawood: any graph of thickness $m$ can be $6m$-coloured. Putting $m = 2$, we deduce that any graph of thickness 2 can be 12-coloured. That is, each node can be assigned one of 12 colours so that nodes that are joined by an edge always have different colours.

So Heawood's theorem implies that the net graph of any PCB can be 12-coloured. We can transfer this colouring (conceptually) to the nets on the PCB. So the nets can each be assigned one of 12 colours in such a way that nets of the same colour are never adjacent to each other.

Since we are seeking short circuits that link adjacent nets, we can therefore restrict our search to short circuits between nets of different colours. Moreover, to discover whether such a short circuit exists, we can lump all the nets of each colour together, in the following sense. For each of the 12 colours we construct a 'probe'. This is a treelike structure made from conducting material that connects all the nets of a given colour together when it is brought into contact with the board (Figure 29) by the test device. Suppose that we choose two colours – say red and green. We attach both the red and the green probes to the PCB, keeping them separate so that no electrical current can pass from the red probe to the green one *except* perhaps along the conducting tracks of the PCB. Now we connect a battery and a light bulb across the two probes, and see whether any current flows.

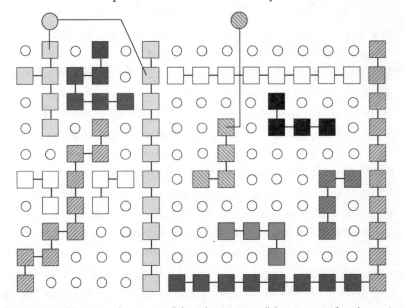

**Figure 29** Attaching two probes: one to all the red nets, one to all the green nets (here just one).

If the PCB has been correctly made, no current will flow, because the red probe connects only to red nets, the green probe connects only to green nets, and on the PCB no red net should connect to any green net. However, if the PCB contains a fabrication fault that links a red net to a green one, then current will flow between the two probes. Now, any fabrication fault in the PCB necessarily connects two adjacent nets, and these must have different colours. So when the PCB is tested using the corresponding two probes, a current will flow in the test device.

Notice that this test doesn't tell us where the error is. But since we are discarding all faulty PCBs, not repairing them, we don't need to know that. The upshot is that in order to detect the presence of a fabrication fault, it is enough to check for the existence of electrical connections – through the conducting material of the board – between all possible pairs of *probes*. Since there are only 12 probes, the number of such pairs is $12 \times 11/2 = 66$. So instead of 125,000 or more tests, we need only 66 – already a major improvement.

However, we can easily do better (Figure 30). Test probe 1 against probe 2; throw out any PCBs with connections between them. Now add a 'gate' to connect probes 1 and 2. Test probe 3 to see if it connects to the circuit formed by probes 1, 2, and the gate. If so, then probe 3 connects either to probe 1 or to probe 2. Either eventuality is a mistake, so we don't care which one occurs: we just throw the PCB out. Now add a second gate connecting probe 3 to the previous two, and continue in this manner. That gets the number of checks down to 11.

Allen Schwenk (West Michigan University, Kalamazoo) realized that a further reduction can be made. Write the numbers 1, . . ., 12 in binary: 0001 up to 1100. Make a 'superprobe' that connects all probes that start with 0; make another that connects those starting with 1. Test whether these two superprobes are connected. If so, throw out the PCB. If not, create two more superprobes connecting probes that have the same binary digit in the second place. Check whether these are connected. Do the same for the third place and the fourth place in the binary expression. That's it. To see why it works, note that if two distinct

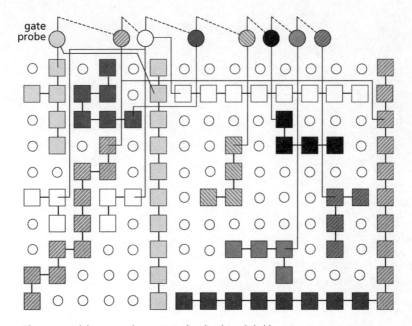

**Figure 30** Joining a complete system of probes by switchable gates.

probes are connected by a short circuit, then their binary expressions must differ in at least one of the four places, so one or other of the four tests will detect the mistake.

Of course there may be other errors in the PCB, but the ones eliminated by this method are much the most common. And a reduction from 125,000 tests *per board* to only four is well worth having as soon as the production run becomes reasonably big – because you only need build those complicated probes and superprobes *once* for each design of PCB. Indeed, a suitable 'programmable' probe/superprobe unit could cover all eventualities.

We started out, one chapter ago, with a recreational puzzle about colouring maps in Earth/Moon empires. Now we've ended with a money-saving test technique for PCB manufacturers. What matters in mathematics is not the particular realization of an idea, but what that idea opens up when you pursue it with skill and imagination.

# 11

# Resurrection Shuffle

Shuffle the cards, shuffle the cards, shuffle the cards, shuffle the cards, shuffle the cards, shuffle the cards, shuffle the cards, shuffle the cards, shuffle the cards, shuffle the cards . . . Oops, we're back where we started. It's the poker-player's nightmare, come to life. Number theory explains why it happens.

# M

**ost card games begin with** someone shuffling the deck. The aim of shuffling, of course, is to randomize the order in which the cards appear . . . but some methods for shuffling cards can achieve the exact opposite. If the cards are shuffled *too* perfectly, using a systematic method, the results can be very far from random. Stage magicians exploit this effect in some of their tricks; card players may wish to avoid it.

As an example, we'll look at one common type of shuffle – or, rather, two very closely related variants – and see what they can be made to do. Specifically, we will analyse the 'riffle' shuffle, in which the deck is divided into two equal parts, which are then interlaced alternately. American magicians call this the Faro shuffle, and to English magicians it is the weave shuffle. Since the two parts into which the deck is separated are of equal sizes, the number of cards in the deck has to be even. (It is possible to consider an analogous shuffle with an odd number of cards, in which one part has one card more than the other, but for simplicity I will ignore this possibility.)

Let's see what effect a riffle has. A full pack of 52 cards is a bit complicated, so to begin with we'll suppose that the deck has ten cards, numbered 0–9. Number them so that initially all cards are face down and the deck is arranged in numerical order from the top down, like this:

0 1 2 3 4 5 6 7 8 9

To perform a riffle, cut the deck between 4 and 5 and interlace. We get either

0 5 1 6 2 7 3 8 4 9

if the top card is taken from the top half of the deck, or

5 0 6 1 7 2 8 3 9 4

if the top card is taken from the bottom half of the deck. The first method is called an *out* shuffle, the second an *in* shuffle.

The theory of in and out shuffles was studied in depth by Persi Diaconis (Stanford), Ron Graham (Bell Labs), and Bill Kantor (University of Oregon) in *Advances in Applied Mathematics* in 1983. They also compiled some information on the history of card shuffles. The earliest recorded mention of the riffle shuffle that they found dates from 1726, in a book called *Whole Art and Mystery of Modern Gaming*, author unknown. In 1843 J. H. Green described the riffle shuffle to Americans in *An Exposure of the Arts and Miseries of Gambling*, showing how it could be used to cheat at the game of Faro. Magicians learned of the shuffle from C. T. Jordan's *Thirty Card Mysteries* of 1919. An early riffle-shuffler, the Nebraskan rancher Fred Black, used to practise the technique on horseback, and he worked out a lot of the mathematics of repeated out shuffles of the standard 52-card deck. Many of the main theorems for decks of any size were published by Alex Elmsley, a computer scientist living in London, in 1957. Some of his results were anticipated by the French mathematician Paul Levy in the 1940s, and further results were proved by Solomon Golomb, the inventor of the famous pentominoes™ puzzle, in 1961.

The analysis can be reduced purely to in shuffles, which simplifies the description considerably, if we are willing to reduce (conceptually) the number of cards in the deck by two. Specifically, an out shuffle can be viewed as an in shuffle on a deck with two fewer cards – just remove the top and bottom cards from the original deck.

To see how this reduction works, consider the ten-card deck above. If we take the cards 0 and 9 in the original order, marking all but the end cards in bold type, we get

0 **1 2 3 4 5 6 7 8** 9

The result of an out shuffle is

0 **5 1 6 2 7 3 8 4** 9

and we see that the cards excluding 0 and 9, shown in bold type, have been subjected to an in shuffle, while 0 and 9 have not moved.

By the reverse reasoning, an in shuffle can be converted to an out shuffle by adding two extra cards to the pack, one at the top and one at the bottom. For many purposes this connection allows us to consider just one of the two shuffles, and we'll focus on the in shuffle. The main question of interest in this chapter is: what happens to the pack if we use an in shuffle several times in a row? Does the deck just get more and more jumbled?

Let's see what happens with the ten-card deck. Here are the results of the first few shuffles:

Shuffle 0: 0 1 2 3 4 5 6 7 8 9
Shuffle 1: 5 0 6 1 7 2 8 3 9 4
Shuffle 2: 2 5 8 0 3 6 9 1 4 7
Shuffle 3: 6 2 9 5 1 8 4 0 7 3
Shuffle 4: 8 6 4 2 0 9 7 5 3 1
Shuffle 5: 9 8 7 6 5 4 3 2 1 0

So although at first the ordering seems to get more jumbled, by the fifth shuffle the entire deck has exactly reversed its order! Clearly five more riffle shuffles will reverse the order again, 'resurrecting' the original order. We conclude that the in shuffle, applied repeatedly to ten cards, cycles repeatedly through just ten different orders. This is a tiny fraction of the 3,628,800 different ways to place ten cards in order.

It's a coincidence that the number of repeats needed to get back to the starting order is ten, the same as the number of cards; but it's not a coincidence that *some* number of repeats restores the original order.

If you try the same kind of calculation with different (even) sizes of decks, you will find that the deck *always* returns to its original order if the in shuffle is repeated sufficiently many times. It's not at all obvious how many times, though: it depends on the number of cards in a rather erratic way.

First, let's see why enough repetitions of the in shuffle restore the order. Figure 31 shows how each card moves when an in shuffle is applied. For example card 0's place is taken by card 5, card 1's place by card 0, and so on. Following the arrows, we see that the cards take each other's places in the following order:

$$0 \to 5 \to 2 \to 6 \to 8 \to 9 \to 4 \to 7 \to 3 \to 1 \to$$

repeating again from 0. The ten cards form a single 'cycle', and with each riffle the cards move one step further along this cycle. Since the cycle contains ten cards, we see that after ten riffles every card returns to its starting point.

The main atypical feature of this deck is that there is just one such cycle. A more typical case is the eight-card deck (Figure 32). Now there are two cycles:

$$0 \to 4 \to 6 \to 7 \to 3 \to 1 \to$$

repeating from 0, and

$$2 \to 5 \to$$

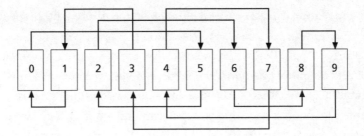

**Figure 31** How an in shuffle cycles cards in a deck of 10.

repeating from 2. The first cycle repeats after six steps, the second after two. When the first cycle has reached its first repeat, after six steps, the second cycle has repeated for the third time. That is, after six steps, *both* cycles repeat. So with a deck of eight cards, the entire pack returns to its original order after six repetitions of the in shuffle.

However many cards there are, and whatever *fixed* rule is used to shuffle them, the progress of the cards through the deck can be broken down into a number of such cycles. Why? Start from any card and follow its progress. Since the deck is finite, eventually the card must reach a position it has previously occupied. From that stage on it will repeat its previous moves. A cycle, however, should repeat from the *beginning*. At the moment, for all we know a card might move something like this:

$$0 \rightarrow 5 \rightarrow 2 \rightarrow 6 \rightarrow 8 \rightarrow 2 \rightarrow 6 \rightarrow 8 \rightarrow 2 \rightarrow 6 \rightarrow \ldots$$

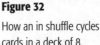

**Figure 32**

How an in shuffle cycles cards in a deck of 8.

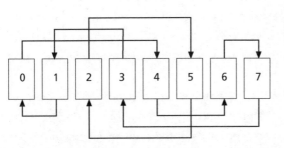

with a repeating cycle $2 \to 6 \to 8 \to$ stuck on the end of a non-repeating bit $0 \to 5 \to$, say.

Can we be sure that when the card first repeats a previous position, it repeats its *initial* position? The answer is yes, and the reason is that any shuffle is reversible – it can be 'undone' by shuffling the card 'back in time'. If the first repeat was not the initial position then we could backtrack one step to find an earlier repeat. For similar reasons, a cycle cannot 'run into' another one. So every card occurs in exactly one cycle.

Once we know the cycles, there is a simple way to work out how many shuffles it takes for the entire deck to be 'resurrected' into its original order. The cycles have various lengths, and each card in that cycle repeats its position after a number of shuffles equal to that length. Suppose, for the sake of argument, that the cycles have lengths 3, 5, and 7. The first cycle repeats whenever the number of shuffles is divisible by 3. The second cycle repeats whenever the number of shuffles is divisible by 5. The third cycle repeats whenever the number of shuffles is divisible by 7. So in order for all three cycles to repeat, the number of shuffles has to be divisible by each of 3, 5, and 7. The smallest such number is $3 \times 5 \times 7 = 105$, obtained by multiplying all the cycle lengths together.

This rule holds however many cycles there are – that is, however many cards we have, provided it is finite. Sometimes a repetition occurs sooner – for instance, take the eight-card deck. There the cycles have lengths 2 and 6, but the order of the cards repeats after six shuffles. It does also repeat after $2 \times 6 = 12$ shuffles, but that's not the smallest number that works. In general, the smallest number of shuffles needed for a repeat can be found by forming the lowest common multiple of the cycle lengths; that is, the smallest number divisible by all of them. All cards will return to their initial position after that number of shuffles.

For example, suppose the cycles have length 10 and 14. The cards in the cycle of length 10 return to their original positions at stages 10, 20, 30, 40, 50, 60, 70, and so on. The cards in the cycle of length 14, on the other hand, return to their original positions at stages 14, 28, 42, 56, 70, and so on. The first number common to both sets, the lowest common multiple of 10 and 14, is 70. And on the 70th shuffle *all* the cards return to their original places.

The in shuffle, then, always repeats, no matter how big the deck may be. But the number of times required for a repeat has no obvious pattern – the fact that a ten-card deck takes ten shuffles to repeat, equal to the size of the deck, is *not* typical. In fact decks of 4, 6, 8, 10, 12, 14, 16, 18, 20, 22, and 24 cards respectively require 4, 3, 6, 10, 12, 4, 8, 18, 6, 11, and 20 in shuffles, respectively, to bring them back to their original order.

Although there is no *obvious* pattern, there is still a pattern. You just have to be a number theorist to spot it! It works like this. Take the case of 8 cards. Add one to the size of the pack to get 9. Form successive powers of 2, divide by 9, and work out the remainders:

| power | 1 | 2 | 3 | 4 | 5 | 6 |
|---|---|---|---|---|---|---|
| value | 2 | 4 | 8 | 16 | 32 | 64 |
| remainder | 2 | 4 | 8 | 7 | 5 | 1 |

The remainder equals 1 for the sixth power – and the number of riffles needed to 'resurrect' an eight-card pack is 6. Similarly when there are 10 cards we add 1 to get 11, and consider the remainders on dividing the powers of 2 by 11:

| power | 1 | 2 | 3 | 4 | 5 | 6 | 7 | 8 | 9 | 10 |
|---|---|---|---|---|---|---|---|---|---|---|
| value | 2 | 4 | 8 | 16 | 32 | 64 | 128 | 256 | 512 | 1024 |
| remainder | 2 | 4 | 8 | 5 | 10 | 9 | 7 | 3 | 6 | 1 |

We get remainder 1 for the 10th power, and that is the correct number of riffles to resurrect a ten-card deck.

This rule works in general. You don't actually need to perform the calculation in such a laborious way: instead you just start with 2, repeatedly double and find the remainder on dividing by one more than the size of the deck, and keep going until you hit 1. A general result in number theory called Fermat's Little Theorem – discovered by the great French number theorist Pierre de Fermat, more famous for his 'Last Theorem' recently proved so gloriously by Andrew Wiles (Princeton University) – implies that this process reaches 1 after a number of steps that is at most the size of the deck.

Because an out shuffle is the same as an in shuffle on a deck with two fewer cards, a similar rule holds for the out shuffle, but now you *subtract* one from the size of the deck and find remainders upon dividing powers of 2 by that. For a standard 52-card deck, the relevant numbers are 52 for the in shuffle, but only 8 for the out shuffle.

In *Mathematical Carnival* Martin Gardner suggests a practical method for testing out such results, which is to *work backwards*. It is difficult, even for an expert magician, to perform an exact riffle shuffle once, let alone repeatedly. But working backwards is easy: deal the deck as if to two people and then stack their hands on top of each other. The reverse of an in shuffle is called an *in sort*, and the reverse of an out shuffle is called an *out sort*. The number of steps needed to resurrect the original order of the deck is the same whether you shuffle or sort.

Many card tricks exploit regularities of the riffle shuffle. Gardner's column in the August 1998 issue of *Scientific American* included a riffle-shuffle trick that works even if you perform the riffle badly! Here's one involving an odd number of cards – though it starts with a deck of 20. Hand this deck to your victim, turn your back, and ask him to shuffle them (by any method), then to insert the joker and remember the two cards it went between. Turn round, and take the deck – which now has 21 cards – face down. Perform either an in sort or an out sort and let the victim cut the deck; do this again. Open the cards in a fan, holding them so your victim can see their faces but you can't, and ask him to take out the joker. Break the fan at that point, fold up each part and put the two together *swapping their order*. Perform two out sorts and an in sort, put the deck face down on the table. Ask the victim to name the two cards he remembered. Turn over the top card: it will be one of them. Turn over the entire pack: the bottom card is the other.

The most difficult question answered in the work of Diaconis, Graham, and Kantor is this: what rearrangements of a deck of $2n$ cards can be obtained by using arbitrary sequences of in shuffles and out shuffles? The results depend on $n$ in a very curious way. If $n$ is a power of 2, the number of such rearrangements is relatively small ($k2^k$ if $n = 2^k$). Otherwise, the number of rearrangements is quite a bit larger, but still falls short of the full $(2n)!$ possibilities. The exact number depends on whether $n$ is of the form $4m$, $4m + 1$, $4m + 2$, or $4m + 3$ for integer $m$. Moreover, the cases $n = 6$ and $n = 12$ are exceptional and do not follow the otherwise general pattern. Sorry, mathematics is often like that – even when there is a pattern, it may split into several parts and there may be some exceptions, usually early on. If you want to see the details, read their beautiful paper.

# Double Bubble, Toil and Trouble

Every physicist in the world knows what shape two bubbles form when they join together. So does every child who has ever blown soap bubbles. Every mathematician in the world knows what shape two bubbles *ought* to form when they join together. A few very clever mathematicians have now *proved* that everyone else is right.

The dodecahedron, a familiar mathematical shape, has 20 vertices, 30 edges, and 12 faces – each with 5 sides (Figure 33). But what solid has 22.83 vertices, 34.25 edges, and 13.42 faces – each with 5.103 sides? Some kind of elaborate fractal, perhaps? After all, fractals – those complex shapes that Benoit Mandelbrot turned into a comprehensive theory of nature's irregularities – can have non-integer dimensions, so why not non-integer vertices? No, this solid is an ordinary, familiar shape, one that you can probably find in your own home. Look out for it when you drink a glass of cola or beer, take a shower, or wash the dishes.

I've cheated, of course. My bizarre solid can be found in the typical home in much the same manner that 2.3 children can be found in the typical family. It exists not as a single thing, but as an average. And it's not a solid, it's a bubble – the 'average' bubble in a mass of foam. Foams contain thousands of bubbles, crowded together like tiny, irregular polyhedrons – and the average number of vertices in these foam

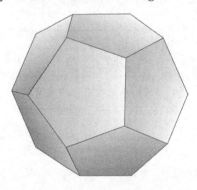

**Figure 33**
The dodecahedron.

polyhedrons is 22.9, the average number of edges is 34.14, and the average number of faces is 13.39. If the average bubble did exist, it would be like a dodecahedron, only slightly more so.

Bubbles have fascinated people ever since the invention of soap; foams have been around since the dawn of time. But the mathematics of bubbles and foams only really got going in the 1830s, when the Belgian physicist Joseph Plateau began dipping wire frames into soap solution and was astounded by the results. Despite 170 years of research, we still do not have complete mathematical explanations – or even descriptions – of many of the phenomena that Plateau observed. A notorious case, until recently, was the Double Bubble Conjecture, which describes the shape formed when two bubbles coalesce. Everyone 'knows' that it should look like Figure 34a – but what about Figure 34b, for instance? Why can't that happen?

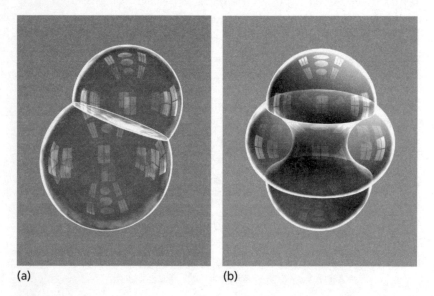

(a)                                    (b)

**Figure 34**

(a) The Double Bubble Conjecture states that when two bubbles coalesce they form two spheres, meeting at 120° along a spherical boundary.

(b) Other possibilities that must be ruled out include this peanut-in-a-tube.

Many other phenomena found by Plateau, however, are now well understood, and experiments with soap films have repeatedly played a role in helping mathematicians to develop rigorous proofs of important geometric theorems. When Plateau began his work on bubbles he was losing his sight. In 1829 he had carried out an optical experiment that involved looking at the sun for 25 seconds: this damaged his eyes, and by 1843 he was totally blind. But his loss of vision did not prevent him making major contributions to that most intensely visual area of mathematics, three-dimensional geometry. Indeed, his work in that area continued long after he had lost all vestiges of sight.

Soap bubbles and films are examples of an immensely important mathematical idea, called a 'minimal surface'. This is a surface whose area is the smallest possible, subject to satisfying certain additional constraints.

Minimal surfaces turn up in the mathematics of bubbles because of a physical effect known as surface tension. The surface of a liquid behaves *as if* it were elastic, like a thin skin of rubber. If you try to stretch the surface, a force opposes it. The force is caused by the structure of the molecules at the surface, which differ from those in the interior of the fluid because some of their chemical bonds are, in effect, missing. The result of surface tension is to store energy in the surface.

The mathematics of missing chemical bonds is distinctly complicated, but fortunately there is a simple approximation that is extremely accurate if all we are interested in is the overall shape of the surface, not the molecular details. It turns out that the energy due to surface tension in a soap film is proportional to its area.

A soap bubble is a minimal surface – that is, a minimal-*area* surface – because it is 'really' a minimal-energy surface. Since energy is equal to area for surface tension (well, they are proportional, which is the same up to some constant factor), minimizing area is the same as minimizing energy. As it happens, nature likes to minimize energy – so bubbles minimize area.

For example, it can be proved mathematically that the surface of smallest area that encloses a given volume is a sphere – and that's why soap bubbles are spherical. A soap bubble encloses a fixed volume of air, and a soap film is so thin – around one millionth of a metre – that it closely resembles an infinitely thin mathematical surface. (Moving bubbles are another matter, because dynamical forces can make them wobble into all sorts of fantastic shapes.) There are many applications of minimal surfaces, including biology, chemistry, crystallography, even architecture.

Without some constraint, the area of a minimal surface would be zero – which is, after all, as small as *any* area can get. The commonest constraints are that the surface should enclose some given volume, that its boundary should lie on some given surface, that its boundary should be some curve, or some combination of these. A bubble that forms against a flat table top, for example, is usually a hemisphere, and this is the smallest area surface that encloses a given volume *and* has a boundary lying in a plane (the top of the table).

Plateau was especially interested in surfaces whose boundary was some chosen curve. In his experiments, the curve was represented by a length of wire, bent into shape, or several wires joined together in a frame. What, for example, is the shape of a minimal surface whose boundary comprises two identical 'parallel' circles? A first guess might well be that it is a cylinder. However, we can do better. Leonhard Euler proved that the true minimal surface with such a boundary is a catenoid (Figure 35), formed by revolving a U-shaped curve known as a catenary about an axis running through the centres of the two circles. The catenary is the shape formed by a heavy, uniform chain hanging under gravity: it looks rather like a parabola, but has a slightly fatter shape. (A hoary mathematical joke goes 'how do you make a catenoid?' 'By pulling its tail.' Its effectiveness depends on how you pronounce 'catenoid'.) Euler's theorem can be demonstrated by making two circular wire rings, with handles – like fishing net frames. Hold them together, dip them into a bowl of soap solution

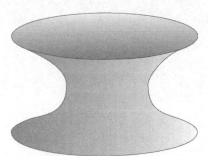

**Figure 35**
A catenoid – the minimal surface spanning two parallel circles.

or detergent, and gently pull them apart to reveal the catenoid in all its glistening beauty.

One of the most famous descriptions of the mathematics of soap films can be found in the classic *What Is Mathematics?* by Richard Courant and Herbert Robbins. They relate some of Plateau's original experiments, in which he dipped wire frames shaped like regular polyhedra. The simplest case, which they don't discuss, arises when the frame is a tetrahedron, a shape with four triangular sides and six equal edges. Here the minimal spanning surface consists of six triangles, all meeting at the centre of the tetrahedron (Figure 36a). A cubic frame leads to a more complicated arrangement of 13 *nearly* flat surfaces (Figure 36b). The tetrahedron case is fully understood by mathematicians, but a complete analysis for the cube remains elusive.

The tetrahedral frame illustrates two important general features of soap films, observed empirically by Plateau. Along the lines running from the vertices of the frame to its central point, soap films meet in threes, at angles of 120°; at the central point, four edges meet at angles of 109° 28'. These two angles are fundamental to any problem in which several soap films abut each other. Angles of 120° between faces and 109° 28' between edges arise not just in the regular tetrahedron, but in any arrangement of soap films whatsoever – provided there is no trapped air, or, if there is, that the pressures on the two sides of each film are equal, and hence cancel each other out.

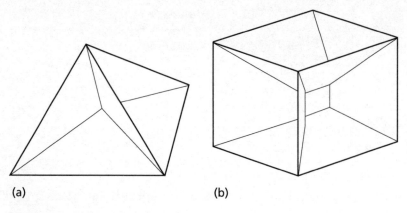

(a)                                     (b)

**Figure 36**

(a) Soap film in a tetrahedral frame forms six planar surfaces.

(b) Soap film in a cubic frame forms 13 almost planar surfaces.

The films in a foam are slightly curved, but can be approximated by plane faces: with this approximation, the two stated angles will be observed in the interior of a foam, though not for films near the foam's external surfaces. This fact is the basis of a curious calculation, which leads to the strange numbers with which I began this chapter. By pretending that a foam is made from many identical polyhedrons whose faces are regular polygons with angles of 109° 28′ (which is impossible, but who cares?), we can estimate the average numbers of vertices, edges, and faces in any foam (see the box).

Plateau's observation about the 120° angle was quickly established as a mathematical fact. The proof is often credited to the great geometer Jacob Steiner, in 1837, but Steiner was beaten to the punch by Evangelista Torricelli and Francesco Cavalieri around 1640. All these mathematicians actually studied an analogous problem for triangles. Given a triangle and a point inside it, draw the three lines from that point to the triangle's vertices, and add up their lengths. Which point makes this total distance smallest? Answer: the point that makes the three lines meet at angles of 120°. (Provided no angle of the triangle

●●●●●●●●●●●●●●●●●●●●●●●●●●●●●●●●●●●●●●●●●●●●●●●●●●●●●●●●●●●●●●●

## SMARTER THAN THE AVERAGE FOAM

Assume that the bubbles in a foam are regular polyhedrons, whose faces are regular polygons with $n$ sides, and that the angles between these sides are all $A = 109° 28'$. Since no such object exists, let us call it the 'follyhedron', and pretend that it does anyway. Let the follyhedron have $V$ vertices, $F$ faces, and $E$ edges.

It is well known that for a regular polygon with $n$ sides and angle $A$ (measured in degrees) we must have $n = 360/(180 - A)$. (For example, if the angle is $90°$, then $n = 360/90 = 4$, a square, as you'd expect.) The reason is that there are $n$ external angles of $180 - A$, which must add up to a full $360°$. With $A = 109° 28'$, this equation implies that the follyhedron has $n = 5.104$ sides.

From here on out the calculation gets slightly more complicated. At each vertex of the follyhedron, three faces meet – because $A$ is bigger than $90°$ but less than $120°$. So the total angle at each vertex is $3A$. Summing over all vertices, we get a total angle of $3VA$. However, the same sum can be found by summing over all faces, each of which contributes $nA$ to the total angle. Therefore $3VA = nFA$, so $3V = nF = 5.103F$, whence

(1)   $V = 1.701F$.

Now consider the $E$ edges. Each face has $n$ edges, giving $nF$ edges in total. But each edge is common to two faces, so in fact

(2)   $E = nF/2 = 2.552F$.

Finally recall Euler's famous formula

(3)   $F + V - E = 2$,

which is valid for any polyhedron. Using (1) and (2) to replace $V$ and $E$ in (3) by multiples of $F$, we get $F + 1.701F - 2.552 F = 2$, which simplifies to give $0.149F = 2$, so $F = 2/0.149 = 13.42$. Then $V = 22.83$, and $E = 34.25$.

●●●●●●●●●●●●●●●●●●●●●●●●●●●●●●●●●●●●●●●●●●●●●●●●●●●●●●●●●●●●●●●

exceeds 120°, that is: otherwise the desired point is the corresponding vertex.) The problem for soap films can be reduced to that for triangles by intersecting the films with a suitable plane.

In 1976 Frederick Almgren and Jean Taylor proved Plateau's second rule about 109° 28' angles. Their ingenious proof ran through a series of steps. They started by considering any vertex where six faces meet along four common edges. First, they showed that the slight curvature that occurs in most soap films can be ignored, so that the films can be considered planar. They then considered the system of circular arcs formed by these planes when they intersect a small sphere centred on that vertex. Because the soap films are minimal surfaces, these arcs are 'minimal curves' – their total length is as small as possible. By the spherical analogue of the Torricelli–Cavalieri theorem, these arcs must always meet in threes, at angles of 120°. Almgren and Taylor proved that exactly ten distinct configurations of arcs – they are rather complicated so I won't draw them – can satisfy this criterion. For each case, they asked whether the total area of the films inside the sphere could be made smaller by deforming the surfaces slightly, perhaps introducing new bits of film. Any such cases could be discarded, since they could not correspond to a true minimal surface. Exactly three cases survived this treatment. The corresponding arrangements of film are a single film, three meeting along an edge at 120°, or six meeting at 109° 28' – just as Plateau observed. The detailed techniques required for the proof went beyond geometry into analysis – calculus and its more esoteric descendants. Almgren and Taylor used abstract concepts known as 'measures' to allow their proof to contemplate bubble shapes far more complex than smooth surfaces.

The 120° rule leads to a beautiful property of two coalescing bubbles. It has long been assumed on empirical grounds that when two bubbles stick together, they form three spherical surfaces, arranged as in Figure 37. If so, the radii of the spherical surfaces must satisfy a beautiful relationship. Let the radii of the two bubbles be $r$ and $s$, and let the radius of the surface along which they meet be $t$: then the relationship is $1/r = 1/s + 1/t$. This fact is proved in Cyril Isenberg's delightful book *The Science of Soap Films and Soap Bubbles*, using no more than elementary geometry and the 120° property.

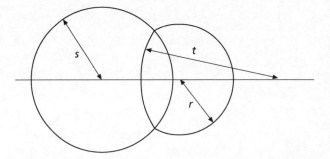

**Figure 37** Conjectured geometry of a double bubble, shown in cross section. Rotate the arcs about the horizontal straight line to obtain the surfaces. The radii $r$, $s$, $t$ satisfy the relation $1/r = 1/s + 1/t$.

All that remains is to prove that the surfaces are parts of spheres, and it is this apparently obvious step that caused all the trouble. In 1995 Joel Hass (University of California, Davis) and Roger Schlafly (Real Software, Santa Cruz) obtained a proof – but only by making the additional assumption that the bubbles are of equal volume. Their proof required the assistance of a computer, which had to work out 200,260 integrals associated with competing possibilities – a task that took the machine a mere 20 minutes!

It took another five years of head-scratching before the complete solution was found. In 2000, the double bubble conjecture was proved for bubbles with unequal volumes, by Michael Hutchings (then at Stanford, now at Berkeley), Frank Morgan (Williams College), Manuel Ritoré (Granada), and Antonio Ros (Granada).

Bubbles continue to pose challenging problems for mathematicians. We now know a lot more than Plateau did when he first dipped his wire frames into sudsy water, but we should also remember that it was those experiments that created a beautiful area of mathematics: the geometry of minimal surfaces.

# 13

# Crossed Lines in the Brick Factory

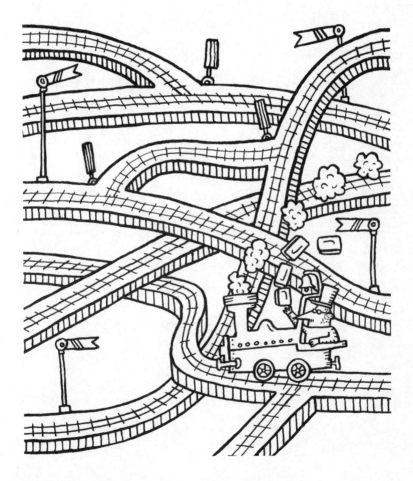

Towards the end of the Second
World War, a Hungarian
mathematician was working in a
brick factory, and he noticed that the
little trains that carried the bricks kept
falling off the lines where they
crossed. An engineer would have
redesigned the lines.
Guess what the mathematician did.

Oneofthecharmsofmathematics is how problems with very simple ingredients, ones that are easy to state and are consistent with a lot of evidence, can baffle the best brains in the world for centuries. Examples include Fermat's Last Theorem, the Kepler Problem, and the Four-Colour Conjecture – all of which have been solved within the past few decades, as I shall briefly describe in a moment. One of the delights of recreational mathematics is the possibility, however unlikely, of finding a solution to some famous unsolved problem. The Four-Colour Conjecture in particular attracted a lot of attention from recreational mathematicians, and it was in some ways a pity when it was proved, because a source of apparently endless fun had dried up. Given all the recent progress, it might seem that there are no interesting challenges left for the amateur to have a go at – but this is not the case, as we'll see.

First, a few words about those three big problems. Around 1650 Pierre de Fermat wrote in the margin of a textbook that he had proved that it is impossible for two perfect cubes to add up to another perfect cube, and that the same is true for fourth powers, fifth powers . . . any power higher than the square. Despite numerous attempts at a proof, this theorem remained open until Andrew Wiles of Princeton University finally polished it off in 1996. The tale was the subject of a prizewinning television programme. Earlier, in 1611, Johannes Kepler wrote (in a New Year's gift to his sponsor, his delightful book *On the Six-Cornered Snowflake*) that he was convinced that the most efficient way to pack spheres in three-dimensional space is the arrangement that many greengrocers employ to stack oranges – a series of honeycomb-like layers stacked on top of each other so that each layer fits into the indentations in the one below. A computer-assisted proof of this was announced in 1998 by Thomas Hales, and has since been published. The Four-Colour Conjecture, roughly a hundred years old, asked whether every map in the plane can be coloured using at most four colours so that adjacent regions are assigned different colours. It was proved by Kenneth Appel and Wolfgang Haken in 1976, again with computer assistance.

The Four-Colour *Theorem*, as it now is, belongs to the area of mathematics known as graph theory. Recall that a graph is a collection of 'nodes', represented by dots, joined by 'edges', represented by lines. A map in the plane, and the notion of 'adjacent region', can be encoded as a graph. There is one node for each region, and an edge joins nodes for which the corresponding regions are adjacent. So the Four-Colour problem can be rephrased as a question about colouring the nodes of a suitable graph.

Graph theory is a source of numerous problems that are simple to state and tricky to answer. Many of these are still open, and one area that includes many such questions concerns the *crossing number* of a graph. Draw the graph in the plane (on a sheet of paper, if you wish) in such a way that the number of crossings of edges is as small as possible. (Edges are not permitted to meet nodes except at their ends, and edges should cross each other at isolated points.) This minimal number of crossings is, of course, the aforementioned crossing number. Nadine Myers of Hamline University discussed this question in *Mathematics Magazine* in 1998. She quoted a remark made by Paul Erdös and Richard K. Guy in 1970: 'Almost all questions one can ask about crossing numbers remain unsolved.' That remark is equally true today. In fact it is astonishing how little is known about the crossing number.

Although it seems very hard to *prove* much about crossing numbers, recreational mathematicians can get a lot of pleasure by experimenting with diagrams of graphs and trying to reduce the number of crossings. It is conceivable that such experimentation might even *disprove* some outstanding conjectures, by reducing the crossing number to less than the conjectured value.

Graphs with crossing number *zero* have been fully characterized, a result dating from 1930 and known as Kuratowski's Theorem, after Kazimierz Kuratowski who first proved it. Such a graph is *planar* – it can be drawn in the plane without any crossings. The graph in Figure 38a is in fact planar. Although it is drawn with four crossings, the edges and nodes can be moved around to get rid of all crossings, as in Figure 38b. In fact this graph is just a 'cycle' on six nodes (six nodes joined in a ring). Similar graphs with $n$ nodes can be defined, and are denoted by the symbol $C_n$. So this graph is $C_6$.

Kuratowski's Theorem states that a graph is planar if it does not contain (in a slightly technical sense) either of the graphs shown in Figure 39a and b. (Note that along the edge of these graphs there may occur nodes that subdivide those edges.) Figure 39a (ignoring such extra nodes) is called the 'complete graph with five nodes': every node is joined to every other. There are analogous complete graphs with any number of nodes. If there are $n$ nodes, this graph is denoted by $K_n$. We met it in Chapter 9. As a reminder, Figure 39a shows $K_5$. Figure 39b (again ignoring extra nodes) is the 'complete bipartite graph on two sets of three nodes'. The nodes fall into two sets, each containing three nodes, and each node in one set is joined to all the nodes of the other set. Similar

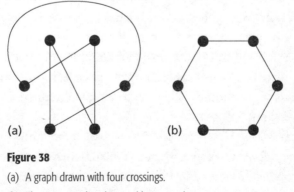

**Figure 38**

(a)  A graph drawn with four crossings.

(b)  The same graph redrawn with no crossings.

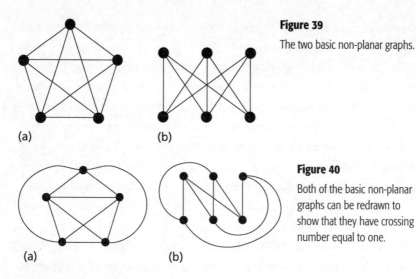

**Figure 39**
The two basic non-planar graphs.

(a)　　　(b)

**Figure 40**
Both of the basic non-planar graphs can be redrawn to show that they have crossing number equal to one.

(a)　　　(b)

graphs can be defined when the two sets of nodes have other numbers of elements, not necessarily equal. If there are $m$ nodes in one set and $n$ in the other, this graph is denoted by $K_{m,n}$. The figure shows $K_{3,3}$.

Both $K_5$ and $K_{3,3}$ have crossing number 1. Given that they are not planar, this can be seen by redrawing the edges to avoid each other whenever possible, and observing that you get just one crossing. See Figure 40a, b.

The concept of crossing number seems to have arisen in 1944, during the Second World War, when the Hungarian mathematician Paul Turán was working in a brick factory outside Budapest. The factory had a number of kilns, where the bricks were baked, and a number of storage yards. Railroad tracks ran from each kiln to each yard. Workers put bricks on to a small truck and pushed it along the rails to a yard; then unloaded the truck. This was a relatively easy task except where one set of rails crossed another. Then the truck often jumped the rails and the bricks fell out.

An engineer would probably have considered how to redesign the crossings. Turán, being a mathematician, wondered how to create as few crossings as possible by redesigning the layout of the rails. After a few days he realized that in this particular factory there were unnecessary crossings, but the general problem intrigued him. With $m$ kilns and $n$ storage yards, and assuming that every kiln has rails to every yard, the problem is: find the crossing number of the complete bipartite graph $K_{m,n}$.

A fair amount is known about graphs with very small crossing numbers $(0, 1, 2)$. Very little is known about graphs with larger crossing numbers. In fact the only cases of this kind where the crossing number is known are $K_n$ for $n \leq 10$, $K_{m,n}$ for $3 \leq m \leq 6$, and graphs $C_m \times C_n$, defined in a moment, for $3 \leq m \leq 6$ and $m = n = 7$.

The graphs $C_m \times C_n$ arise from 'rectangular grids on a torus'. Figure 41 shows an example, $C_7 \times C_8$. I have drawn the graph as two families of circles. The 'concentric' circles form 7 copies of $C_8$, and the 'radial' circles (drawn as ellipses) form 8 copies of $C_7$. These circles can be drawn on the surface of a torus, and on the torus they cross only at the black dots. But when the diagram is projected into the plane, additional crossings occur. In fact here there are 5 such crossings for each of the 8 vertical circles, a total of 40.

The same kind of construction can be carried out with $m$ horizontal

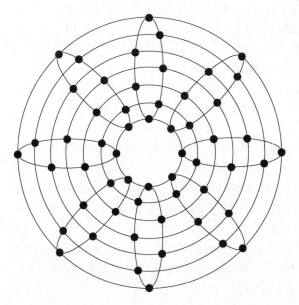

**Figure 41**

The torus-grid graph $C_7 \times C_8$, here drawn with 40 crossings. Can this number be reduced?

circles and $n$ vertical ones, where we make the convention that $m \leq n$. Then each vertical circle crosses all but two of the horizontal circles *twice*. It meets the others – the 'inner' and 'outer' horizontal circles – once, at a node. For the other circles, one such crossing is a real crossing on the torus, hence a node; the other, however, is a result of trying to draw the picture in the plane. So each vertical circle contributes $m - 2$ crossings. In total, therefore, we find $(m - 2)n$ crossings.

It is widely believed that this is the minimal number; that is, that the crossing number of the torus-grid graph $C_m \times C_n$ is $(m - 2)n$. However, this '$(m,n)$-conjecture' has never been proved. It *is* known to be true for the cases listed above, the most recent being $C_7 \times C_7$. (See Myers's article for the details and references.) The smallest unsolved case is therefore $C_7 \times C_8$, for which the conjectured number of crossings is 40.

Can you find a way to redraw Figure 41 in the plane with 39 or fewer crossings? No cheating, no clever 'cookery' of the problem, please! This is mathematics, not a puzzle. If so, the $(m,n)$-conjecture would be *false*. Experiment.

It may seem astonishing that the combined brainpower of the world's mathematicians can't determine whether Figure 41 can be redrawn with fewer crossings – but it provides graphic illustration (pun intended) of the difference between a question that is simple to ask and one that is simple to answer.

Even if improvements are possible, they ought to be minor. In 1997 G. Salazar (Carleton University) proved that if the crossing number of $C_m \times C_n$ is *less* than $(m - 2)n$, then it cannot be *much* less. Assuming one technical condition (the number of times any two $n$-cycles cross cannot exceed some stated bound), the crossing number divided by $(m - 2)n$ approaches 1 as $n$ becomes arbitrarily large. Nevertheless, this result leaves room for a reduction in the conjectured value $(m - 2)n$ for any particular choice of $m$, $n$. If the conjecture is false, that would help explain why it seems so hard to prove. On the other hand, it might be like Fermat's Last Theorem, the Kepler Problem, and the Four-Colour Conjecture: *true* but hard to prove!

# 14

# Division without Envy

Whichever way you cut it, the issue keeps coming back to how to divide the cake fairly, equitably, while respecting human rights and providing equal opportunities for all. This chapter is committed to fair shares for every citizen, irrespective of colour, creed, gender, age, or mathematical orientation.

So why are you *still* not satisfied?

**I**n the opening chapter we took a look at some of the mathematical issues arising from the deceptively simple problem of dividing a cake fairly – meaning that if there are $n$ people then each is convinced that their share is at least $1/n$ of the cake. Now we'll pick up the story and take a look at some of the more modern parts of the theory.

A brief reminder of where we'd got. With two people, the time-honoured algorithm 'I cut, you choose' leads to fair division. With three or more people, there are several possibilities. The 'trimming' method allows successive participants to reduce the size of a purportedly fair share of the cake, with the proviso that if nobody else trims that piece, then the last person to trim it has to accept it. In the 'successive pairs' algorithm, the first two people divide the cake equally, and then the third person secures what they consider to be at least $1/3$ of each piece by negotiating with each of the first two separately. And with the 'divide and conquer' algorithm, participants try to divide the cake using one cut so that roughly half the people would be happy to have a fair share of one piece, while the rest would be happy to have a fair share of the other piece. The same idea is then repeated on the two separate subcakes, and so on.

These algorithms are all fair, but there is a more subtle issue. Even if everybody is convinced that they have a fair share of the cake, some participants may still feel hard done by, thanks to the Deadly Sin of envy. For example, Tom, Dick, and Harry may all be satisfied that they've got at least $1/3$ of the cake; nevertheless, Tom may feel that Dick's piece is bigger than his. Tom's share is 'fair', but he doesn't feel quite so happy any more. A division of the cake is 'envy-free' if no person thinks that someone else has got a larger piece than they have. An envy-free division is always fair, but a fair division need not be envy-free. So finding an algorithm for envy-free division is a more difficult problem than finding a fair one.

Cut-and-choose for two people is easily seen to be envy-free, but none of the other algorithms mentioned above is – not for all possible sets of three valuations of the cake. An envy-free algorithm for three people was first found by John Selfridge and John Conway in the early 1960s:

> STEP 1: Tom cuts the cake into three pieces that he considers to be of equal value.
>
> STEP 2: Dick may either (a) do nothing, if he thinks that two or more pieces are tied for largest, or (b) trim what he perceives to be the largest piece to create such a tie. Set aside any trimmings: call the accumulated trimmings 'leftovers'.
>
> STEP 3: Harry, Dick, and Tom, in that order, choose a piece – one that they consider to be either the largest or tied for largest. If Dick trimmed a piece in step 2 then he must choose the trimmed piece *unless* Harry has already done so.

At this stage part of the cake has been divided in an envy-free manner. So it remains to divide up the leftovers in an envy-free manner too.

> STEP 4: If Dick did nothing at step 2 there are no leftovers and the cake has been divided. If not, either Dick or Harry took the trimmed piece. Suppose Dick took it (if Harry did, interchange those two people from now on in the description of what to do). Then Dick divides the leftovers into three pieces that he considers equal.

STEP 5: All that remains is for Harry, Tom, and Dick to choose one piece from the leftovers, in that order. Then Harry has first choice so has no reason to be envious. Tom will not envy Harry *however* the leftovers are divided, because the most that Harry can get is a piece that Tom is already convinced is worth 1/3. And he won't envy Dick because he chooses before Dick does. Dick has no grounds for complaint since it was he who divided the leftovers anyway.

At this point everyone got stuck for thirty years. Is there an envy-free protocol for *n* people? In 1995 Steven Brams and Alan Taylor discovered a remarkable envy-free protocol for any number of players. It is distinctly complicated and I won't give it here: see either their article in the *American Mathematical Monthly* or the marvellous book *Cake Cutting Algorithms* by Jack Robertson and William Webb, both listed in Further Reading.

What other related questions are there? One possibility is to assign *unequal* shares. For example, Alice and Bob may want to divide the cake so that Alice is convinced she gets 3/5 or more, while Bob is convinced he gets 2/5 or more – that is, both seek a ratio of 3:2. It turns out that this problem has very different solutions depending on whether the desired ratio is expressible in whole numbers, or whether it is an irrational ratio like $\sqrt{2}:1$. In the former case, Alice could be replaced by three clones and Bob by two, who then divide the cake fairly. In the latter case, this approach doesn't work because you can't make $\sqrt{2}$ clones of somebody. Nonetheless, in the irrational case the division can still be achieved in a finite number of cuts, though you can't predict in advance how many will be needed.

One of the most interesting features of the theory of cake division is what Robertson and Webb call the 'serendipity of disagreement'. At first sight it might seem that fair division is simplest when everybody is in agreement about what each bit of the cake is worth – after all, there can then be no disputes about the value of a given share. Actually, the reverse is true: as soon as participants disagree about values, it becomes *easier* to keep them all happy.

Suppose, for example, that Tom and Dick are using the cut-and-choose algorithm. Tom cuts the cake into two pieces, which he views as having equal value, $1/2$ each. If Dick agrees with those valuations, nothing more can be done. But suppose that Dick values the two pieces at $3/5$ and $2/5$. Then he might, for some altruistic reason, decide to give Tom $1/12$ of what he considers to be the larger piece (which he values at $1/20$ of the whole cake). He still has $3/5 - 1/20 = 11/20$ of the cake, according to his valuation. One way to do this is for Dick to divide the larger piece, in his estimation, into 12 parts that he considers of equal value. Then he offers Tom the choice of just one of them. Whichever one Tom chooses, Dick still thinks he is left with $11/20$. Tom, on the other hand, is faced with 12 choices, and he values their total as $1/2$. Therefore at least one of them is worth $1/24$ in his estimation. By choosing that one, he ends up with what he considers to be at least $13/24$ of the cake. So now both Tom and Dick are satisfied that they have *more* than a fair share.

The intuition here is not that disagreement about values must lead to disagreement about what constitutes fair division. That might happen if a third party divided the cake and then insisted that Tom and Dick accepted one of those predetermined shares, but it can easily be avoided if Tom and Dick do the dividing themselves. Because in that case, if Tom places more value on a piece than Dick does, then Tom will be *more* easily satisfied. The trick is to make the cuts and choices in the right manner, that's all. There is a message for political disputes here: a solution is more easily found if the parties concerned can be brought to the negotiating table to thrash out the deal *themselves*. A deal imposed by an outside body, however fair it may seem to be to a disinterested observer, may not be acceptable to the actual participants.

Another instance of the same principle arises in the problem of dividing beachfront property. Suppose that a straight road runs east–west past a lake, and that the land in between the road and the lake is to be divided by north–south property lines. The problem is to divide the property among $n$ people, so that each of them gets a connected plot of land that they consider to be at least $1/n$ of the total value. The solution is disarmingly simple. Make an aerial photograph of the property and ask each participant to draw north–south lines on it, so that in their estimation the land is divided into $n$ plots of equal value (Figure 42). If all participants draw their lines in the same places, then any allocation will satisfy them all. If there is any disagreement at all about where the lines go, however, then it is possible to satisfy all of them that they have a fair share, *and* to have some of the property left over. Figure 43 shows a typical case where Tom, Dick, and Harry have carried out such a procedure. Clearly we can let Tom have his first plot, Dick his second, and Harry his third – with some bits left over. Figure 44 shows a more complicated example in which Tom, Dick, Harry, Marcia, and Becky each seek $1/5$ of the land. In 1969 Hugo Steinhaus proved that the same thing happens for any choices of dividing lines where there is the slightest disagreement. A proof, using the principle of mathematical induction, can be found in the book by Robertson and Webb.

You might like to consider whether a similar method would work with a cake. Ask each player to mark radial lines on a photograph of the cake, dividing it into what they consider to be pieces worth $1/n$. Then compare their choices. It's a very similar problem but with one

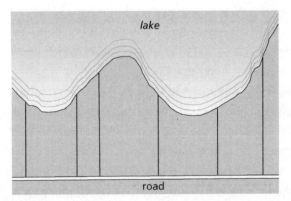

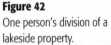

**Figure 42**
One person's division of a lakeside property.

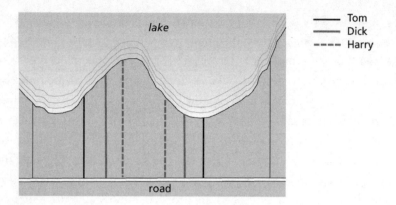

**Figure 43** Keeping three people happy, with land to spare.

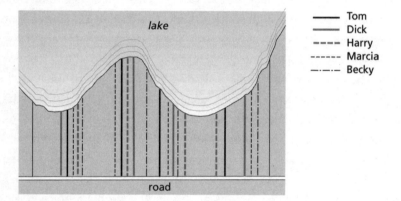

**Figure 44** Keeping five people happy, with land to spare.

snag: the cake 'wraps round' into a full circle. But what if you begin by marking a single radial line, the same one on every photo, and insist that they make one of their cuts along that line?

Disagreements in valuation can be made to work the other way, too. Sometimes people want not the biggest share, but the smallest. For example, can Tom and Dick divide the task of mowing a lawn so that they each think their piece is less than half the lawn? This is the 'dirty work' problem, a relatively neglected relative of the cake-cutting problem. You might like to consider modifying the algorithms for fair division of the cake so that each of $n$ persons mowing a lawn considers that their bit is *at most* $1/n$ of the lawn.

Sadly, not all chores can be divided fairly, at least, not with reasonable restrictions. Take washing-up. If each person must wash and/or dry a complete dish, then in extreme cases no fair allocation is possible. Imagine two participants with one huge dish and one small one. Both will want to deal with the small one, and won't accept the huge one. So even in a perfect world, where all disputes are settled by negotiation, some disagreements seem unavoidable.

## FEEDBACK

James Fradgley sent an amusing comment on the realities that underlie the issue of cake division. I reproduce it in its entirety:

A delightful mathematical approach, but it simply doesn't work, as many individuals have an 'other man's grass is always greener' approach to life. So what may seem fair at one time, may not a few minutes later. When my children were aged about 4 and 5, my wife divided a small cake and gave each of them an estimated 50% share. My daughter (the older) immediately said: 'His is larger than mine.' So my wife asked our son if he thought his was larger than his sister's. He replied he didn't think so, and agreed that he was happy if they were swapped. My wife then swapped them, in the simple belief that both would be happy.

*But . . .*

Our daughter looked at the swapped plates, and said 'His is *still* larger than mine.' Alas, division without envy is nothing to do with reality or calculation!

# Furiously Flashing Fireflies

When male fireflies in some parts of Asia go on the pull, they join forces to attract the opposite sex, and entire trees light up with simultaneous green flashes as they signal their desire for a mate. We can all understand *why* they do that – but *how* do they do it?

O ne of the most spectacular displays in the whole of nature occurs soon after sunset in South-East Asia, where huge swarms of fireflies flash in synchrony. As the American biologist Hugh Smith wrote in 1935:

> Imagine a tree thirty-five to forty feet high, apparently with a firefly on every leaf, and all the fireflies flashing in perfect unison at the rate of about three times in two seconds, the tree being in complete darkness between flashes. Imagine a tenth of a mile of river front with an unbroken line of mangrove trees with fireflies on every leaf flashing in synchronism, the insects on the trees at the ends of the line acting in perfect unison with those between. Then, if one's imagination is sufficiently vivid, he may form some conception of this amazing spectacle.

Why do the flashes synchronize? The biological reason seems to be evolutionary. The flashes are created solely by male fireflies, and they attract females. Synchronized flashes attract them from further away, offering an evolutionary advantage.

What about the mathematical reason? In 1990 Renato Mirollo and Steven Strogatz showed that synchrony must inevitably occur in certain mathematical models, which assume that every firefly interacts with every other in a particular manner. Their idea is to model the insects, and the signals that pass between them, as a population of mathematical oscillators, coupled together by visual signals. The model incorporates some key biological features of real fireflies, but of course it is a simplification. I'll explain the word 'oscillator' in a moment.

Fireflies employ a special light-emitting chemical to create a flash. They have a good supply of the chemical, but they release it in small bursts according to a repeating cycle of 'readiness'. In effect, it is as if the fly starts counting steadily from zero as soon as it has flashed, and only when it reaches a hundred does it flash again. Its state of readiness – the number its count has reached, so to speak – is the 'phase' of the cycle.

Mathematically, such a cycle is an oscillator – a unit whose natural dynamic causes it to repeat the same behaviour over and over again. The population of fireflies can be represented by a network of such oscillators that are 'coupled together' – interact – in a 'fully symmetric' manner. That is, each oscillator affects all of the others in exactly the same manner. The most unusual feature of this model, which was first introduced by the physiologist Charles Peskin in 1975, is that the oscillators are 'pulse-coupled'. This means that an oscillator affects its neighbours only at the instant when it creates a flash of light. The mathematical difficulty is to disentangle all of these interactions. Mirollo and Strogatz do this by applying techniques from dynamical systems theory, in which oscillators are an especially important component.

Oscillators are a source of periodic rhythms, which are common – and important – in biology. Our hearts and lungs follow rhythmic cycles whose timing is adapted to our body's needs. Many of nature's rhythms are like the heartbeat: they take care of themselves, running 'in the background'. Others are like breathing: there is a simple 'default' pattern that operates as long as nothing unusual is happening, but there is also a more sophisticated control mechanism that can kick in when necessary and adapt those rhythms to immediate needs.

Why do systems oscillate? Oscillation is the simplest thing you can do if you don't want, or are not allowed, to remain still. Why does a caged tiger pace up and down? Its motion results from a combination of two constraints. First, it feels restless and does not wish to sit still.

Second, it is confined within the cage and cannot simply disappear over the nearest hill. The simplest thing you can do when you have to move but can't escape altogether is to oscillate. Of course there is nothing that forces the oscillation to repeat a regular rhythm; the tiger is free to follow an irregular path round the cage. But the simplest option – and therefore the one that is most likely to arise both in mathematics and in nature – is to find some series of motions that works, and repeat it over and over again. And that is what we mean by a periodic oscillation. A more physical example is the vibration of a violin string. That, too, moves in a periodic oscillation; and it does so for the same reasons as the tiger. It can't remain still because it has been plucked away from its natural resting point; and it can't get away altogether because its ends are pinned down.

The oscillations of fireflies are created by a mechanism known as 'integrate-and-fire' – or, in their case, 'integrate-and-flash'. In such oscillators some quantity builds up ('integrates') until it reaches a threshold. Crossing the threshold value triggers a sudden change ('fire' or 'flash') in which the quantity is reset to zero, after which it starts to build up again (Figure 45). In the firefly, this quantity is the phase of the cycle that determines when to release a burst of the chemical that causes a flash. When the phase reaches threshold, the fly flashes; the phase resets to zero and the process starts again. Observations in the laboratory and in the wild show that when other fireflies notice the flash they get excited, and their own phase receives a sudden boost. This moves them nearer to threshold.

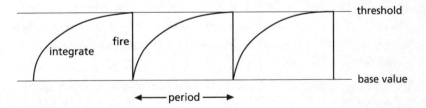

**Figure 45** Integrate-and-fire oscillator.

Oscillators are said to be 'coupled' if one affects the state of the other. The classic example is the observation, made by the great Dutch physicist Christiaan Huygens, that pendulum clocks sitting on the same shelf affect each other. Each pendulum makes the shelf vibrate, and the vibrations are transmitted to the other pendulum. Often this interaction causes the pendulums to synchronize.

However, coupled oscillators do not *always* synchronize, a good example being an animal's legs when it walks. Each leg is an oscillator, and the animal's body couples them, but the legs do not normally all move at once. A swarm of fireflies behaves like a system of coupled oscillators, and it seems that for this system, synchrony is the norm. The mathematician's task is to work out why.

Peskin took the first step towards such an understanding. He introduced a specific model of an integrate-and-fire oscillator in a study of the synchronization of the muscle fibres of the heart. His model gives a specific equation for how the phase builds up. This same equation can be applied to fireflies – physiological studies show that it is a reasonable, though not exact, representation of the flash cycle. Peskin also introduced the important idea of pulse coupling for integrate-and-fire oscillators. Here an oscillator affects the others *only* when it fires. Then it sends a signal to the others, which gives their phases a boost. If this boost tips another oscillator over the threshold, then it, too, fires, and so on.

It turns out that the chemicals in fireflies are affected in just this manner by visual signals from other fireflies. When a firefly sees another one flash, it gets excited, and this moves it closer to threshold. Peskin proved that if two identical pulse-coupled integrate-and-fire oscillators obey his equation, then they will eventually synchronize. (Actually, if their initial phases are set to very special values, their flashes will alternate periodically, but this state is unstable – it can be destroyed by the smallest disturbance. Apart from these special values, the system always synchronizes. So we say that it 'almost always' synchronizes.)

He also conjectured that the same would be true of any network of coupled integrate-and-fire oscillators. Mirollo and Strogatz proved this conjecture, assuming a more general equation than Peskin's. Subject to a few technical hypotheses, stated in their article, they showed that in a system with any number of identical integrate-and-fire pulse-coupled oscillators and all-to-all coupling, then almost always the oscillators eventually become synchronized. (Again, there is a rare set of initial conditions in which the behaviour is periodic, but again these states are unstable, hence the 'almost'.) Their proof is based on an idea called 'absorption', which happens when two oscillators with different phases lock together and thereafter stay in phase with each other. Because the coupling is fully symmetric, once a group of oscillators has locked together it cannot unlock. A geometric and analytic proof shows that a sequence of these absorptions must eventually lock all of the oscillators together.

We can explore the firefly system with a more simplified model – the solo game of Flash, played with counters moving round the edge of a square. I'll illustrate Flash with a $6 \times 6$ square, but you can use any size you like – an $8 \times 8$ chessboard or a $10 \times 10$ Monopoly™ board works well. Flash uses only the outer border (Figure 46). One corner square (bold edges) is singled out as the 'flash' square. The four edges are assigned numbers 1, 2, 3, 4 in order, going clockwise. A few counters representing fireflies are placed at random: I've shown three but you can use any number you wish. The position of a firefly indicates its phase: the further clockwise the firefly is, the closer it is to threshold. The 'flash' square is the threshold value at which the firefly flashes and resets its supply of chemical to zero.

The game proceeds in a series of 'stages' in which each firefly moves at least once. The rules for each stage are:

1. Move each firefly one square clockwise (increment the phase according to its natural cycle). To interpret the rules it helps to think of them as moving simultaneously, though in practice you have to move them one at a time.

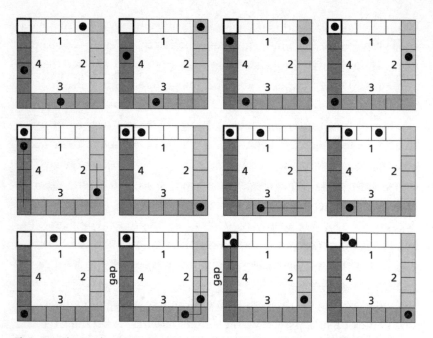

**Figure 46** The opening stages of a game of Flash. 'Gap' indicates that some stages are omitted. Lines indicate boosts caused by another firefly flashing.

2. If any firefly has landed on the 'flash' square, move every other firefly clockwise by a number of squares that is equal to the number of the side of the board upon which it sits. For instance, a fly on side 3 moves three squares clockwise. (This is the pulse-coupling. The other flies notice the one that flashes, and move closer to threshold. Fireflies with a bigger phase move more squares, which is how real fireflies behave.)

3. If, during step 2, any firefly would pass through the 'flash' square, stop it exactly on that square.

4. If any firefly lands on the 'flash' square as a result of steps 2 and 3, go back to step 2 for that firefly and again move all the others according to rule 2.

5. If two or more fireflies are on the same square, move them as a single unit and pretend that it is a single firefly.

Figure 46 shows the first few stages of such a game. To save space, the first eight moves are all shown, but after that moves are missed out ('gap') unless a firefly hits the 'flash' square.

In the sequence shown, two of the fireflies eventually get onto the same square, meaning that they have synchronized their flashes. This is a case of 'absorption', and the rules imply that thereafter they move as a unit, so they can never become de-synchronized. If you keep playing, you'll find that eventually all three fireflies synchronize.

Try Flash with different initial placements and different numbers of fireflies. They nearly always synchronize if you keep playing long enough. However, I suspect that for some sizes of boards it may be possible to find initial placements that lead to periodic, asynchronous behaviour. Such placements correspond to the unstable states in the Mirollo–Strogatz theory. The Flash game is a finite-state model, simpler than the one analysed by Mirollo and Strogatz though similar to it, and it may not behave in exactly the same manner.

Similar ideas apply to many systems other than fireflies. Applications include the pacemaker cells of the heart; networks of neurons in the brain, including those controlling circadian rhythms; the insulin-secreting cells in the pancreas; crickets and katydids that chirp in unison; and groups of women whose menstrual periods become synchronized. And, as a bus driver in a local township pointed out to me in a letter, it is also closely related to the phenomenon whereby you wait ages for a bus and none come . . . until suddenly three come along at once. Synchronized buses!

## FEEDBACK

I asked whether the game of Flash could lead not to complete synchrony, but to a situation in which there is some periodic cycle with not all checkers on the same square. (This does not happen in the standard mathematical model of firefly synchronization, where the 'phase' in the cycle is a continuous variable, but it becomes a possibility in the analogous discrete-state problem, which is the game of Flash. It can also occur in other, similar mathematical models with a continuously varying phase – and so, indeed, can 'chaos'.)

William J. Evans of Irvine, CA has discovered that if the game is played on the perimeter of a 12 × 12 'checkerboard' with five fireflies, there exist initial states that lead to a periodic cycle. The upshot of his analysis is that the position of Figure 47a, with five fireflies all at distinct phases, leads, after 27 moves, to that of Figure 47b; moreover, this second configuration repeats after a further 38 moves – creating a period-38 cycle that continues indefinitely.

Cindy Eisner (Rehov, Israel) went to town on the question. She obtained a complete analysis for all moderate-sized boards, finding the largest group of fireflies for which no pair ever converge (up to 16 × 16 board), the number of initial states for which no pair ever converge (up to 15 × 15), and the number of initial states for which there is no final synchronization (up to 11 × 11). For example, on a 4 × 4 board the largest group of fireflies for which no pair ever converge contains four fireflies, which start at positions 1, 4, 7, 11: the dynamics is a

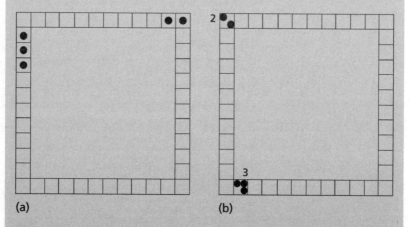

(a)          (b)

**Figure 47** How to obtain states that never synchronize. Start at (a). The game then proceeds to (b), which repeats every 38 moves.

cycle of length 10. On a 15 × 15 board the largest group of fireflies for which no pair ever converge contains 15 fireflies, which start at positions 0, 4, 6, 8, 11, 13, 17, 21, 24, 27, 31, 37, 41, 46, 51: the dynamics is a cycle of length 41. On a 15 × 15 board there are 124,523 initial states for which no pair ever converge, out of a total of $7.20576 \times 10^{16}$ possibilities. On an 11 × 11 board there are $6.76099 \times 10^{10}$ initial states for which there is no final synchronization, out of a total of $1.09951 \times 10^{12}$ possibilities.

Moreover, for any size of board, there are always initial conditions with two fireflies that never converge. For example, put them at positions 0 and $2n - 3$ on an $n \times n$ board. The cycle length is $2n \times 2$, and Cindy conjectured that the states in this cycle are the only non-synchronizing ones for two fireflies.

# Why Phone Cords
# Get Tangled

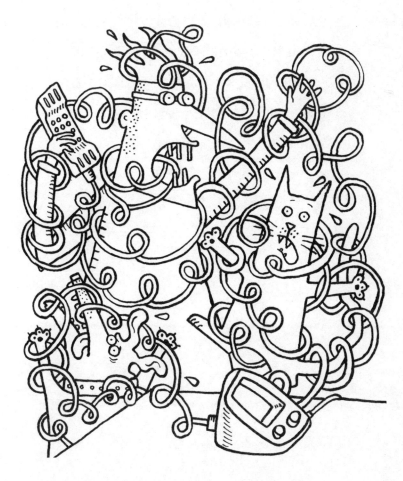

The problem of tangled phone cords is so widespread that companies sell devices to untangle cords, or to stop cords tangling, or to eliminate cords altogether. (It's called 'wireless', which when I was a kid meant 'radio'. How times change.) Why do they tangle, and what has that got to do with DNA?

**W**hy does the phone cord always get twisted?

I'm thinking of those stretchable cords that are formed as a long helical coil, attached to phones that hang on the wall. When you first install the phone, the cord hangs nice and neatly. But as the weeks pass, it gets sort of tangled round itself. You can see the same effect with an elastic band – one like a flat ribbon is best – if you hold its ends loosely between the thumb and forefinger of each hand, and twiddle your fingers (Figure 48). Or you can start with a length of string between your fingers and thumbs, and roll the ends. This kind of behaviour is called 'supercoiling', and it arises in many areas of science, from undersea communication cables to DNA.

I know why the phone cord supercoils in *my* house. The details may not apply in yours, but the general mechanism probably does. It's the same one that makes the elastic band and the string coil round themselves in that characteristic manner. When the phone rings, I pick it up with my right hand and twist it through about a right angle. In order to speak into it, though, I then transfer the phone to my left hand, which imparts a further two right angles of twist. When I've finished, I use my left hand to hang it on the wall again, imparting a final fourth right angle of twist to the cord. So every time I use the phone, I twist the cord by a full 360° – and in the same direction every time.

If I kept it in my right hand, I might well untwist it when I put the phone back. But that transfer between hands seals the cord's fate. The same kind of thing happens to the electric cable for garden tools. After use, I coil it over my shoulder like a mountaineer's rope. Over time, the

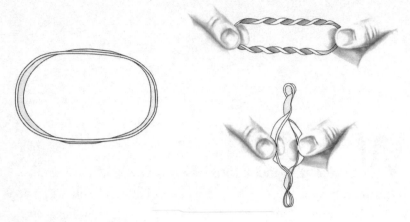

**Figure 48** Supercoiling an elastic band.

cable gets more and more twisted along its length. Something is converting coils into twists – but what?

The branch of mathematics that organizes how we think about this kind of question is topology – 'rubber sheet geometry', the geometry of continuous transformations. Topologists distinguish two different ways to put loops into flat strips: *twists* and *writhes*. To understand both the difference between these, and their relation, it helps to make a long strip of strong paper – I suggest 20 cm long and 1 cm wide. It's useful if one side is distinguishable from the other: colour one side red and the other side blue, or use paper whose sides differ to begin with.

Hold the strip flat and pointing directly away from your body, with your left thumb and forefinger holding the near end, and your right thumb and forefinger holding the far end. Fingers on top, thumbs underneath.

Now move your right hand to coil one loop of the strip round your left middle finger (Figure 49a) – you will need to move your right thumb and finger, without actually letting go of the strip, to do this comfortably. It may sound complicated, but it comes completely naturally if you use a real paper strip. Then remove your left middle finger

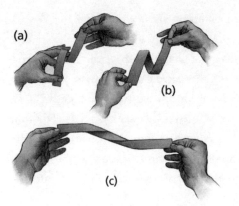

**Figure 49**

(a) Coil a strip around your finger.

(b) Remove finger.

(c) Pull, converting writhe to twist.

to leave a free loop (Figure 49b). If the strip were slightly elastic (as real paper *isn't* but topologists' paper *is*) you could lay the loop out flat, though overlapping itself. In any of these three representations, you have inserted one writhe (coil) into the previously flat loop.

Now, however, go back to the situation of Figure 49b and gently pull your hands apart. The strip deforms into Figure 49c. This is not a writhe, but a twist. You could have got the same effect by holding the strip flat across the front of your body, keeping the left end fixed, and twisting the right end through 360°. So what we see is that one writhe can be deformed, topologically, into one twist.

There's a technical point here that's important. Both writhes and twists have a direction – they can be 'positive' or 'negative'. Deciding which is which is not especially hard, but I don't want to burden you with the details, so you should tackle it in the way that Winnie-the-Pooh tackled the problem of telling left from right. He knew that once he'd worked out which paw was right, then the other was left: the problem was how to begin. Here, once you've decided that a given writhe or twist is positive, then its mirror image is negative. The easy way to start is to declare that the writhe in Figure 49b is positive, but the twist in Figure 49c is negative. So actually the number of twists is *minus* the number of writhes. This choice leads to the simple equation $T + W = 0$, where $T$ is the number of twists and $W$ the number of writhes. With a different convention about signs we'd have $T - W = 0$. Either will do, but you have to choose one and stick with it.

Go back to the start again, with a flat strip, but this time coil the strip twice round your left middle finger – two (positive) writhes. When you pull your hands apart, these turn into a double-twist (through $720°$). So two (positive) writhes can become two (negative) twists. For that matter, two (positive) writhes can also become one (positive) writhe plus one (negative) twist. Experiment with three or four writhes: you'll find that a given number of (positive) writhes can be turned into the same number of (negative) twists.

In fact we can prove this. Figure 50a shows how one positive writhe turns into one negative twist, with the ends always being kept in a fixed orientation – as if you used your fingers to press each end of the strip against the top of a table and then just slid the ends about. Figure 50b shows a series of writhes (here three). These can mentally be subdivided along three 'boundaries' into three separate single writhes. Then each single writhe can be turned into a twist, while keeping the boundary lines flat on the table-top. Because the boundary orientations never change, those three twists naturally 'glue' together into one triple-twist. Clearly there's nothing special about the number three here, so we conclude that a strip with a given number of positive writhes can be deformed into a strip with the same number of negative twists instead. So $T + W = 0$, as I said.

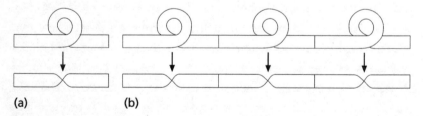

(a)         (b)

**Figure 50**

(a) One writhe becomes one twist.

(b) Repeat the procedure to turn any number of writhes into the same number of twists.

At first sight, a plain string looks different from an elastic band. However, you can keep track of how the string supercoils by imagining that initially a flat band runs along its centre. As you twist one end of the string, this band also twists, and the number of twists in the band counts the number of complete turns you give to the string. If you keep the string taut, all it can do is twist, but if you let the ends move together, the string prefers to writhe, and the supercoil appears.

The reason the string prefers to writhe is related to the fact that it is slightly elastic. Not in the sense of 'elastic band', but in the engineering sense that it is bendy, but produces a restoring force when it bends. The more you bend it, the more strongly it tries to straighten out again. The preference for writhes over twists was first explained in 1883 by A. G. Greenhill, who showed that a writhed shape has less elastic energy than the corresponding twisted one. The same is true even of paper strips, as you can confirm by experiment: unless you impart energy by holding the strip taut, it prefers to writhe. Greenhill added extra detail, proving that if an infinitely long rod is twisted by forces 'at infinity' then it buckles into a helix. In 1990 J. Coyne showed that this helix quickly localizes into a solitary twist, and eventually the rod contracts inwards to turn the twist into a small localized loop – one writhe. If the rod is allowed to contract further, the loop acquires an increasing number of writhes. Recently three Australian mathematicians – D. M. Stump (University of Queensland), W. B. Fraser (University of Sydney), and K. E. Gates (University of Queensland) – analysed the elasticity theory of a twisted rod using more realistic modelling assumptions. They found specific formulas for the exact shape of the supercoil, useful in particular for engineers laying undersea cables, where this kind of twisting is common – and a nuisance.

The situation for the phone cord is in principle more complicated because the cord starts out as a helix anyway – it is already twisted (and/or writhed, depending on your point of view). Nevertheless, a phone cord also converts twists into writhes, just like a plain string – at least if you don't allow its own helical coils to unravel, which is the usual case. (You also get funny 'glitches' in the phone cord where successive coils don't fit together properly – those are more subtle.) You can imagine a long fat string threaded through the helical coils, with a long flat strip embedded in it, and as the cord gets twisted, so does that string, and so does the strip.

The DNA molecule, the hereditary material of living organisms, is – like the phone cord – a helix. More accurately, it is a double helix, in which two helical strands corkscrew round and round each other. Biologists have to understand the geometry of DNA's double helix under a variety of conditions, and they find that it, too, undergoes supercoiling, with transitions from writhes to twists. It is important to understand these transitions when interpreting electron micrographs of loops of DNA (Figure 51). Moreover, as I hinted a moment ago, DNA and phone cords can do something that plain string can't: they can ravel or unravel their own helical coils. One simple topological feature of all this may give you a flavour of the much more sophisticated theories being devised by topologists and biologists. This concerns three features of a closed loop of DNA:

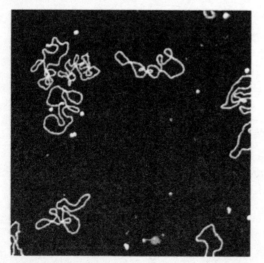

**Figure 51**
Electron micrograph of a DNA loop.

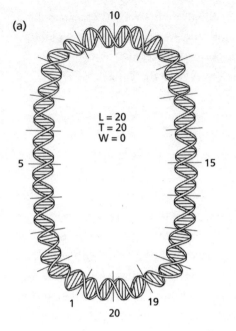

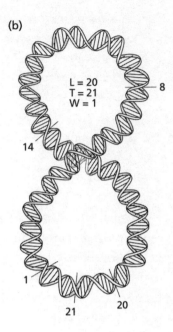

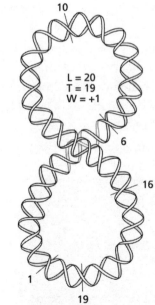

**Figure 52** Trading writhes for twists in a DNA loop.

- The *linking number L*, which is the number of times one strand crosses the other when the molecule is laid out flat in the plane.
- The number $T$ of helical *turns* in the DNA.
- The *writhing number W*, which measures the amount of supercoiling.

The basic formula here is the elegant

$$L = T + W$$

which generalizes our earlier formula '$T + W = 0$' for a flat strip, and can be proved in much the same way. The edges of the flat strip are not linked: for them, $L = 0$. For a given DNA loop, $L$ is fixed, but we can trade writhes for twists or conversely. Figure 52 shows how this works using a DNA loop. The first picture has $L = T = 20$ and $W = 0$. In the second picture, an extra twist is inserted, increasing $T$ to 21. In compensation, a negative writhe ($W = -1$) forms, giving the figure–8 appearance. The third picture shows that if instead we inserted a negative twist (with $T$ becoming 19) then $W$ would change to $+1$. Again we get a figure–8, but the overlap has the other strand on top. For more information see Richard B. Sinden's *DNA Structure and Function*.

There is more, far more, to the topology of DNA than I have room to convey here. But even just the story of writhes and twists provides a fascinating example of the interrelatedness of different aspects of the real world, and the way that simple mathematical principles can reveal that hidden unity.

# Sierpinski's Ubiquitous Gasket

Just over eighty years ago a Polish mathematician invented a curve that crossed itself at every point. Little did he realize that this same shape would turn up all over the mathematical landscape, from Pascal's triangle to the Tower of Hanoi puzzle. But why is the answer 466/885 and not 8/15?

**S**trange numbers, strange shapes . . . these are the things that give mathematics its allure to those who love it. And, even more so, strange connections – topics that seem totally different, yet possess a hidden, secret unity. One of my favourite examples is Sierpinski's gasket, the triangular shape shown in Figure 53. In the term made famous by Benoit Mandelbrot it is a 'fractal', made from smaller copies of itself . . . but it also has connections with self-intersections of curves, Pascal's triangle, the Tower of Hanoi puzzle, and the curious number 466/885, whose numerical value is roughly 0.52655. This number should feature in everybody's list of Numbers that Are More Significant than They Seem, alongside $\pi$, e, the golden number, and the like.

The gasket (the name was supplied by Mandelbrot and is a visual joke) is named after the polish mathematician Waclaw Sierpinski, who was born in Warsaw on 14 March 1882. In 1909 Sierpinski gave the first systematic lecture course ever taught on set theory, and most of his research was in set theory and point set topology. His collected works include an amazing 720 papers (published between 1906 and 1968) plus 106 expository articles and 50 books. He died in Warsaw on 21 October 1969, and his grave bears (in Polish) the well chosen words 'Explorer of the Infinite'.

The gasket made its first appearance in 1915, as 'a curve simultaneously Cantorian and Jordanian, of which every point is a point of ramification'. Less formally, it is a curve that crosses itself at every point – a classic instance of a geometrical property so counterintuitive that such shapes became known as 'pathological curves'. Today they are seen as

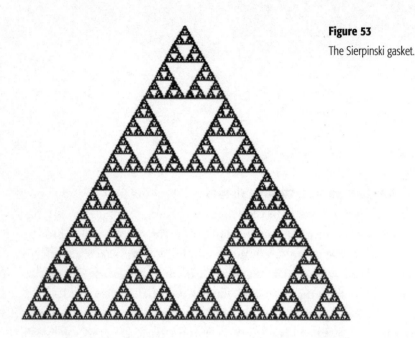

**Figure 53**
The Sierpinski gasket.

a natural and central topic in mathematics, illustrating the dangers of geometric intuition that is too naive, but when they first appeared most mathematicians greeted them as ghastly monstrosities. Sierpinski had more imagination, and found them fascinating.

Strictly speaking, the gasket as drawn crosses itself at every point other than the three corners. Sierpinski's answer to this objection is that if six copies of his triangle are arranged to form a regular hexagon, then the result is a curve that crosses itself at *every* point.

Earlier, in 1890, another Frenchman, Edouard Lucas, discovered a theorem which, in retrospect, provides a connection between the Sierpinski gasket and the celebrated Pascal's triangle (Figure 54) in which each number is the sum of the two above it, immediately to the left and right. These numbers are more technically known as binomial coefficients, and the $k$th entry in row $n$ is the number of different ways to choose $k$ objects out of $n$. The triangle is (mis-)named after Blaise Pascal, who wrote about it in 1665. However, the triangle appears on the title page of an arithmetic text by Petrus Apianus dating from the early sixteenth century, and it can be found in a Chinese mathematics book of 1303. Indeed, it has been traced back at least to Omar Khayyam around 1100, who presumably learned about it from earlier Arabic or Chinese sources. An explicit formula for the numbers in Pascal's triangle was given by Sir Isaac Newton, but in fact this formula was already known to the Indian mathematician Bhaskara in the twelfth century, though not in Newton's notation.

Lucas asked: when is a number in Pascal's triangle even or odd? I shaded Figure 54 to show which entries are odd, but the full pattern requires a larger diagram. It is easy to experiment with a computer: the result, Figure 55, is striking and surprising. The odd binomial coefficients look extraordinarily like a 'discrete' version of the Sierpinski gasket. Lucas found a mathematical explanation for this pattern in 1890, based on the use of binary notation for numbers. Similar patterns can be obtained by asking which binomial coefficients are multiples of 3, or leave remainder 1 or 2 on division by 3 – or, more generally, leave a given remainder on division by some chosen number. The resulting patterns are at least as pretty as that for odd/even: see Marta Sved's article in Further Reading.

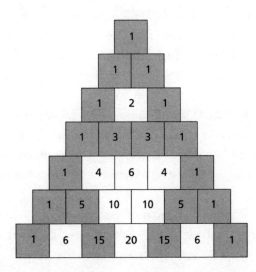

**Figure 54**

Pascal's triangle, with odd numbers shaded.

**Figure 55**
Odd (black) and even (white) numbers in Pascal's triangle.

One curious consequence is that 'almost all' binomial coefficients are even – that is, as the size of Pascal's triangle gets ever larger, the proportion of odd numbers gets closer and closer to 0. The reason is that since the gasket is a curve, its area – which in the limit represents the proportion of odd binomial coefficients – is zero. David Singmaster has taken this observation further, proving that for any $m$, almost all binomial coefficients are divisible by $m$.

Lucas seems to have been haunted – albeit unwittingly – by Sierpinski's gasket. In 1883 he marketed the famous puzzle known as the *Tower of Hanoi*, under the pseudonym N. Claus (the surname being an anagram of his own). It consists of eight (or fewer) discs mounted on three pins – the 3-disc case is shown in Figure 56 – and it is an old friend of recreational mathematicians and computer scientists. The discs are arranged on one pin in order of size, as shown, and they have to be moved one at a time so that no disc ever sits on top of a smaller one, and they all end up in a single pile but on a different pin from the one they start from.

It is well known that the solution has a recursive structure. That is, the solution of $(n + 1)$-disc Hanoi can be simply deduced from that for $n$-disc Hanoi. For instance, suppose you know how to solve 3-disc

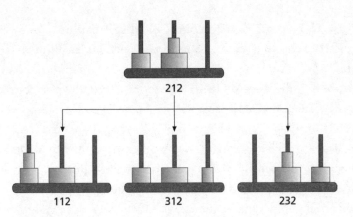

**Figure 56** A typical position in 3-disc Hanoi and the legal moves.

Hanoi, and you are presented with the 4-disc version. Start by ignoring the bottom disc, and use your knowledge of 3-disc Hanoi to transfer the top three discs on to an empty pin. Now, temporarily, notice the fourth disc, sitting all alone on the original pin, and move it to the other empty pin. Now ignore it again, but remember which pin it is now on, and use your knowledge of 3-disc Hanoi to transfer the top three discs on to that pin – on top of the fourth disc, as required. In this manner you can solve 100-disc Hanoi if you know how to solve 99-disc Hanoi, and for the same reason you can solve 99-disc Hanoi if you know how to solve 98-disc Hanoi ... and so on ... which, eventually, gets you down to 1-disc Hanoi. But solving *that* is easy: just pick up the only disc around and move it to any other pin.

We can interpret this recursive structure geometrically, which is where the connection with the gasket comes in. With any puzzle of this general type (moving objects, finite number of positions) we can associate a graph $H_n$, whose nodes are the possible legal positions of the discs, and whose edges represent the legal moves between positions. What does $H_n$ look like? For definiteness, consider $H_3$, which describes the positions and moves in 3-disc Hanoi. To represent a position, number the three discs as 1, 2, 3, with 1 being the smallest and 3 the largest. Number the pins 1, 2, 3 from left to right. Suppose for example that disc 1 is on pin 2, disc 2 on pin 1, and disc 3 on pin 2. Then we have completely determined the position, because the rules imply that disc 3 must be *underneath* disc 1. Thus we can encode this information in the sequence 212, the three digits in turn representing the pins for discs 1, 2, and 3. Therefore each position in 3-disc Hanoi corresponds to a sequence of three digits, each being 1, 2, or 3. There are $3^3 = 27$ positions, because each disc can be on any pin, independently of the others.

What are the permitted moves? The smallest disc on a given pin must be at the top, so it corresponds to the *first* appearance of the number of that pin in the sequence. To move that disc (legally) we must change the number so that it becomes the first appearance of some other number. For example, from position 212 we can make legal moves to 112, 312, and 232, and only these. In this way we can work out all 27 positions and all possible moves, and the graph $H_3$ turns out to be Figure 57. This consists of three copies of a smaller graph (actually $H_2$) linked by three single edges to form a triangle.

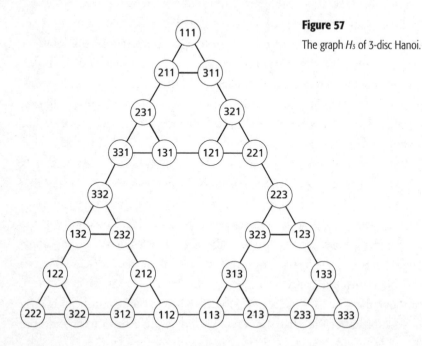

**Figure 57**

The graph $H_3$ of 3-disc Hanoi.

Each smaller graph $H_2$ has a similar triple structure, and this is a consequence of the recursive solution. The edges that join the three $H_2$'s together are the stages at which the bottom disc is moved, and the three copies of $H_2$ are the ways you can move the top two discs *only* – one copy for each possible position of the third disc. The same goes for any $H_n$: it is made from three copies of $H_{n-1}$, linked in a triangular manner. Figure 58 shows $H_5$.

Observe that as the number of discs becomes larger and larger, the graph looks more and more like the Sierpinski gasket.

We can use the graph to answer all sorts of questions about the puzzle. For example, the graph is clearly connected – all in one piece – so we can move from any position to any other. The minimum path from the usual starting position to the usual finishing position runs straight along one edge of the graph, so it has length $2^n - 1$. This result has been known for many years in the form 'the largest disc moves only once'.

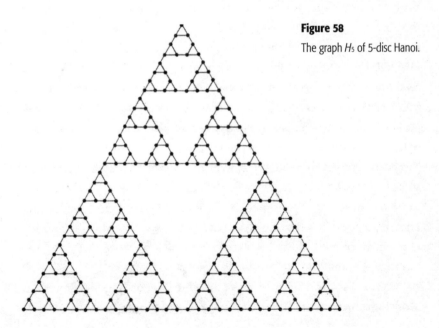

**Figure 58**

The graph $H_5$ of 5-disc Hanoi.

Now for a personal anecdote, which illustrates how ideas can move from recreational mathematics into the mainstream of research. In 1989 I wrote about the graph for the Tower of Hanoi in *Pour la Science*, the French sister magazine of *Scientific American*. Not long after, I attended the International Congress of Mathematicians in Kyoto, and a German mathematician named Andreas Hinz introduced himself. He had been trying to calculate the average distance between two points in a Sierpinski gasket, encountered difficulties, and asked two experts. One said 'It's very difficult' and the other said 'It's trivial and the answer is 8/15.' It turned out that the second expert's proof was fallacious and the first expert was right. The mistake was equivalent to assuming that in the Tower of Hanoi the well known theorem 'the largest disc moves at most once' for the *standard* start and finish positions is still true when moving between *any* two positions by the most efficient route. This is not always true, and it's a common error in the literature.

Unfortunately, even when the nature of the fallacy is grasped, it doesn't help you to find the correct answer. But Hinz had already proved a formula for the average number of moves between positions in the Tower of Hanoi, and using that he could prove that for $n$-disc Hanoi the average number of moves linking two positions is asymptotic to $(466/885)2^n$ – that is, the formula is a good approximation for large $n$. He read my article and realized that his result for $n$-disc Hanoi immediately implies that the average distance between two points in a Sierpinski gasket is precisely $466/885$. (Just divide his formula by $2^n - 1$, the length of the 'side' of $H_n$, and let $n$ become very large. The units of measurement should be chosen to make the side of the gasket have length 1.) This is some 2% smaller than the value 8/15 suggested by the second expert.

At the moment this approach via the Tower of Hanoi is the only known method for finding the answer. For the statistically minded, Hinz also proved that the variance of the distance between two random points in a unit-side Sierpinski gasket is precisely $904808318/14448151575$. Add *that* to your list of Numbers that Are More Significant than They Seem, too!

# FEEDBACK

Ron Menendez of Chatham, NJ pointed out yet another instance of the Sierpinski gasket. Draw three points A, B, C in the plane at the vertices of an equilateral triangle, and pick a random starting point X in the plane. Choose one of A, B, C at random, each with probability 1/3. (For example, roll a die and let 1 or 2 correspond to A, 3 or 4 to B, 5 or 6 to C.) Find the midpoint of the line joining X to the chosen vertex: this is the new position of X. Now repeat, always choosing a random vertex from A, B, C and moving X to the midpoint between its current position and that vertex. Apart from a few initial points where the walk is 'settling down', the resulting cloud of points is – a Sierpinski gasket!

This is quite surprising given the randomness. It is explained by the mathematician Michael Barnsley's theory of self-similar fractals. The Sierpinski gasket has three corners A B C. It is made up from three copies of itself, each half the size: that is, obtained by replacing every point in the gasket with the midpoint of the line joining it to A, to B, or to C. This feature of the gasket corresponds to the rules for the random walk. Barnsley has proved that with probability 1, any random walk following the rules 'converges' to the gasket, meaning that after a few steps every point you draw lies very close to the gasket.

The neat feature of this example is that the gasket emerges in a rather random way from a mist of points, rather than being drawn one piece at a time.

# Defend the Roman Empire!

In the 4th century, the Roman Emperor Constantine lost control of Britain, and shortly afterwards the Roman Empire collapsed completely. It's a pity he didn't know about zero–one programming. All he had to do was work out the best locations for his troops. How many legions should he have sent to Gaul? Egypt? Constantinople? Now we know.

**D**uring the Second World War, when General MacArthur was conducting military operations in the Pacific theatre, he adopted a strategy of 'island-hopping' – moving troops from one island to a nearby one, but only when he could leave behind a large enough garrison to keep the present island secure. As the front of invaded islands advanced, of course, he was able to move troops forward from the rear, so that it was not necessary to retain large garrisons throughout the entire campaign on every island that had been secured.

A similar deployment problem faced the Roman Emperor Constantine in the 4th century, only his task was to maintain the security of the entire Roman Empire. He decided upon what appears to be the first recorded use of the strategy that MacArthur later adopted in the Pacific. In 1997 Charles S. ReVelle (Johns Hopkins University) and Kenneth E. Rosing applied mathematical techniques of 'zero–one programming' to study Constantine's problem and find out whether he might have done better than he actually did. Their work is a beautiful example – simple but instructive – of this technique in action, and it also forms the basis of an enjoyable game. Problems of this type – though usually far more complex – often arise in commercial and military decision-making. An early version of their work was published in *Johns Hopkins Magazine* in 1997, and they presented a more extensive description at ISOLDE 8, the International Symposium on Location Decisions, in 1999.

As a warm-up problem, consider a slightly stripped-down and very simplified version of the Roman Empire at the time of Constantine

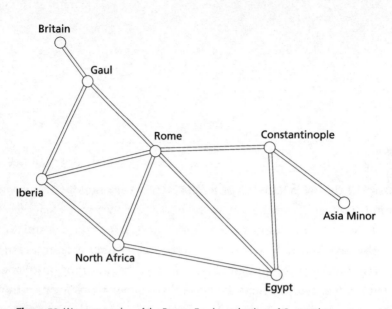

**Figure 59** Warm-up version of the Roman Empire at the time of Constantine.

(Figure 59). This 'game board' shows eight regions, from Asia Minor to Britain, together with routes that link these regions.

A century earlier, the forces of Rome had been dominant throughout most of Europe, and at that time the available troops numbered 50 legions. By the 4th century, however, that number had been halved to 25 legions. Constantine in effect treated these as four groups, each containing six legions, and ignored the single 'spare' legion left over (which in practice made one group contain seven legions, not six). And he devised some simple rules for deploying and moving troops, aimed at producing effective security.

Think of each group of six legions as a single 'piece', to be placed at the circles marked on the game board. Here are Constantine's rules:

- A region is securable if a piece can be moved to it in a single step from one region to an adjacent one.

- However, a piece can be moved in that manner only when a second piece occupies the same region. (Regions can contain as many pieces as you wish – that is, you can station as many legion-groups in any region as you wish.)

Given these rules, how can you allocate your groups to secure the entire empire – or, failing that, as much of it as possible? Constantine's solution was to put two groups at Rome and two at Constantinople. You can check that with this deployment of troops, one region – Britain – is not secured. In fact, with Constantine's rules in force, it takes four moves to get a group to Britain: try to work out a method before reading on.

Here's one way. First, move a piece from Rome to Gaul (thereby securing Gaul, which would doubtless have been a lot more important to the Romans than distant, cold, wet Britain). Then move a piece from Constantinople to Rome, then to Gaul, and finally to Britain itself. Can you find any other method for securing Britain that starts with Constantine's deployment and uses four or fewer moves? If not, can you show that no such methods exist? (When thinking about these questions, it is best to assume that if, say, there are two pieces at Rome, then whichever one of them you choose to move, this represents the 'same' solution.)

Is it possible to improve on Constantine's deployment? Yes, it is, in the sense that a deployment exists in which every region can be secured in at most one move. Again, try to find this before reading on.

In fact, there is exactly one deployment that fits the bill. Namely: place two pieces in Rome, one in Britain, and one in Asia Minor. Why did Constantine not do this? After all, it gives Rome two pieces – 12 legions – just like the emperor's actual solution. It seems likely that he felt unhappy about this solution because it would leave Rome seriously weakened if trouble arose on two different fronts. Once one piece has left Rome, the other is stuck in place – indeed, after that first and only possible move, *all* pieces are stuck.

I said that Figure 59 was a warm-up. The actual problem that faced Constantine is Figure 60, the 'real' Roman Empire, with extra connections between Iberia and Britain, and between Egypt and Asia Minor. Constantine still preferred his deployment, of course. However, we can decide not to worry about second fronts, in which case our 'improved' solution – two pieces in Rome, one in Britain, and one in Asia Minor – still secures the entire empire in at most one move. However, we now have new connections that make further troop movements possible, and we can ask whether there are any other solutions. I will answer that question towards the end of this chapter, but first you might care to experiment – draw the board on a sheet of paper and use coins for pieces. And before I give the answer, let me say a little about the mathematics that can be applied to much more complex problems of the same kind, as well as this one.

The general area here is known as 'programming', and it involves representing all such problems in an algebraic form. One way is to

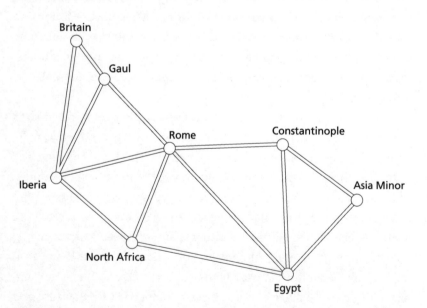

**Figure 60** The actual problem that faced Constantine.

make a table (the fancy term is 'matrix') whose rows correspond to pieces and whose columns correspond to regions. So the matrix has four rows and eight columns. Because here each piece is either in some region or not, we can use a 0 to show that a given piece is not in a given region, and a 1 to show that it is. Figure 61 shows the matrix corresponding to Constantine's actual choice. His rules can all be restated as rules for changing the entries of such matrices, so the problem can be reformulated algebraically. For obvious reasons, questions like this are known as zero–one programming problems.

I won't go into technical details, but it's worth observing that ReVelle and Rosing's method breaks the problem up into two different ones. The first is the Set Covering Deployment Problem (SCDP). This ignores the constraint that there are four pieces, and instead asks for the smallest number of pieces that can be placed so that all regions can be secured in at most one move. (If the answer is 'more than four' then of course the original problem can't be solved.) The second problem is complementary to the first, and is known as the Maximal Covering Deployment Problem (MCDP). This respects the constraint of four pieces, but ignores the need to secure all regions. Instead, it asks for the largest number of regions that can be secured (in one move or none) with four pieces. (Other numbers of pieces can also be considered if necessary.)

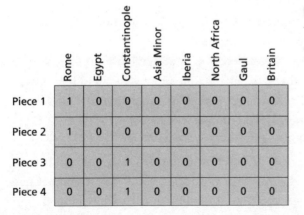

|  | Rome | Egypt | Constantinople | Asia Minor | Iberia | North Africa | Gaul | Britain |
|---|---|---|---|---|---|---|---|---|
| Piece 1 | 1 | 0 | 0 | 0 | 0 | 0 | 0 | 0 |
| Piece 2 | 1 | 0 | 0 | 0 | 0 | 0 | 0 | 0 |
| Piece 3 | 0 | 0 | 1 | 0 | 0 | 0 | 0 | 0 |
| Piece 4 | 0 | 0 | 1 | 0 | 0 | 0 | 0 | 0 |

**Figure 61**

The matrix that Constantine chose.

General methods (embodied as computer software) exist to solve each of these problems, and between them they 'bracket' the original problem, telling us whether a solution exists with four pieces (yes) and whether fewer pieces would work (no). Moreover, the two methods together make it possible to find all possible solutions. Found them yet? This is your last chance, because now I'm going to tell you the answers.

Altogether, there are now six different solutions. The figures in brackets show how many pieces to put in the named regions. (We already know solution 4.)

1. Iberia (2), Egypt (2).
2. Iberia (2), Constantinople (2).
3. Iberia (2), Asia Minor (2).
4. Britain (1), Rome (2), Asia Minor (1).
5. Britain (2), Egypt (2).
6. Gaul (2), Egypt (2).

ReVelle and Rosing's method appears to be the first (and currently the only) method that can solve this kind of allocation problem for a general network. It is practicable for realistically large networks, despite the enormous number of arrangements that could in principle arise.

As it happens, the Emperor Constantine lost control of Britain. The causes were surely more complex than anything that can be captured by this simple model. Nevertheless, it is arguable that if Constantine had been a better mathematician, the Roman Empire might have lasted longer than it did. (I'm joking . . . just. With a more realistic model in a more complex situation, the point may well be valid.)

As it stands, what we have here is a single puzzle, and I've told you the answer. But you can try different networks, different numbers of pieces, and change the rules. In particular, think about competitive versions in which there are two (or more) players, each equipped with their own set of pieces – say red and blue – and pieces are removed from the board if, say, there are more red pieces than blue in a given region. (Here red 'wins' and the blue pieces are captured.) With a bit of experimentation, you can come up with some very playable games.

# Triangulation Takeaway

Brush up your topology with a game whose rules are very simple, but whose strategic considerations are fiendishly difficult. Mathematicians think they know who should win, with perfect play . . . but, as you can guess, they haven't the foggiest idea how to prove they're right.

The tradition of explaining mathematics through games and puzzles goes back at least to the ancient Babylonians, whose clay tablets include puzzles in arithmetic that would be entirely acceptable as 'word problems' today. The rapid growth of whole areas of new mathematics has given rise to entirely new games, whose rules cannot easily be stated without invoking concepts that would have been quite alien to the Babylonians, such as topology or set theory. In an article in the 1996 book *Games of No Chance*, Richard K. Guy (Calgary) reported a game invented by David Gale (Berkeley), which starts out looking like a set-theoretic game and ends up as a topological one. The game holds plenty of interest for recreational mathematicians: for instance, it is still not known which player has a winning strategy, although Gale made a plausible conjecture. Moreover, it is easy to invent variations that are just as much fun to play.

Recall that the basic objects of set theory are *sets*, which are just collections of objects of some specified kind. The objects that belong to a set are its *members*, and are said to be *contained in* that set.

If a set has finitely many members, then we can define it by listing the members inside curly brackets: for example, $\{2,3,5,7\}$ is the set of all prime numbers less than 10. A set $X$ is a *subset* of a set $Y$ if every member of $X$ is a member of $Y$: for example, the set $\{3,5,7\}$ of all odd prime numbers less than 10 is a subset of $\{2,3,5,7\}$. Every set is considered to be a subset of itself; a subset of $X$ is said to be *proper* if it is different from $X$.

Sets can have one member: for example {2}, the set of all even prime numbers. A set can even have *no* members, in which case it is said to be *empty*. An example is the set of all even prime numbers bigger than 3, which in curly bracket form would be { }.

Gale's game is called Subset Takeaway. It starts with a finite set $S$, which we may as well take to be the set {1, 2, . . ., $n$} of whole numbers ranging from 1 up to $n$. Players alternately choose a proper, non-empty subset of $S$, subject to one restriction: no subset chosen earlier (by either player) can be a subset of the new subset. The first player unable to name such a subset loses.

One practical way to play the game is draw up a set of columns on a sheet of paper, headed by the numbers 1, . . ., $n$, and mark a line of crosses in the columns that correspond to the selected subset. A new, legal move cannot include all of the crosses from some previous move. A more interesting way to represent the moves, which we'll come to shortly, is geometric – indeed, topological.

Following tradition, let the players be Alice and Bob, with Alice to move first. When $n = 1$ there are no legal moves. When $n = 2$ we have $S = \{1,2\}$. The only opening moves available to Alice are {1} and {2}, and whichever she chooses, Bob can choose the other. Then Alice is stuck, so Bob wins.

When $n = 3$, we have $S = \{1,2,3\}$. Suppose Alice chooses a subset with two members, say {1,2}. Then Bob can choose the complementary subset (everything not chosen by Alice), which here is {3}. Now Alice can't choose anything that contains 3, so she has to select a subset of {1,2}, and from that point on the entire game is exactly the same as if the starting set had been {1,2}, since Bob can't choose any subset that contains 3 either. So again Bob wins. The same goes if Alice opens with any other two-member subset, for the same reason. However, Alice has another possible kind of opening: a one-member subset, say {3}. Now Bob chooses the complementary subset {1,2}, and again the game must continue as if the starting set had been {1,2}, and Bob still wins. Since Alice's opening must be either a one-member subset or a two-member subset, Bob has a winning strategy: 'always play the complement of Alice's move'.

Before reading on, you may wish to consider whether the same strategy gives Bob a win when $n$ is larger than 3.

Enter the topology. Topology is usually described as 'rubber sheet geometry', the study of properties of shapes that do not change when the shape is continuously deformed. Here, though, we don't need any elastic. Instead, we use one of the basic techniques in topology, which is – when possible – to triangulate the shape: that is, to split it up into triangles that join edge to edge. Strictly speaking, this description applies only to surfaces, but the same approach works for higher-dimensional shapes if we replace triangles by 'simplexes'. A three-dimensional simplex, or 3-simplex, for example, is a tetrahedron, with vertices 1, 2, 3, 4. It has four faces, six edges, and four vertices. The faces are triangles: 2-simplexes. The edges are line segments: 1-simplexes. And the vertices are points: 0-simplexes. Moreover, these bits of the 3-simplex correspond exactly to subsets of {1,2,3,4}. The tetrahedron itself corresponds to the whole set {1,2,3,4}. The faces correspond to the 3-element subsets {1,2,3}, {1,2,4}, {1,3,4}, and {2,3,4}. The edges correspond to the 2-element subsets {1,2}, {1,3}, {1,4}, {2,3}, {2,4}, and {3,4}. And the vertices correspond to the 1-element subsets {1}, {2}, {3}, {4}.

In the same way, an $(n-1)$-simplex can be identified with the set {1, 2, . . ., $n$}, and its various lower-dimensional faces (from now on we use this term irrespective of their dimension) can be identified with subsets whose size exceeds the dimension by 1.

We can now reformulate Subset Takeaway as Simplex Erasure. Players start with a simplex. A move consists of choosing a proper sub-simplex of any dimension, and erasing its interior, as well as all higher-dimensional sub-simplexes that have it as a face. However, the boundary of the chosen sub-simplex – all of its faces – still remains.

We can use this topological representation to analyse Simplex Erasure for a 3-simplex, which corresponds to Subset Takeaway for $n = 4$. The starting position is a complete 3-simplex; that is, a tetrahedron. Because the complete set {1,2,3,4} is not an illegal move, this tetrahedron is 'hollow' – its interior is not available as a move. Figure 62 shows a series of legal moves (diagrams of this type, built up from simplexes of various dimensions, are called *simplicial complexes*). A systematic consideration of all such sequences shows that there is a winning strategy for Bob in the $n = 4$ game. The same goes for $n = 5$, 6, and Gale was led to conjecture that whatever the value of $n$, Bob has a winning strategy. To the best of my knowledge this is not yet proved or disproved.

In 1997 J. Daniel Christiansen (MIT) and Mark Tilford (Caltech) applied more sophisticated ideas from topology to come up with a technique known as 'binary star reduction', which can be used to simplify the game's analysis in certain circumstances. Suppose that at some stage during the game we arrive at a position (represented by a simplicial complex) in which there are two vertices $x$ and $y$ that form a *binary star* – meaning that three conditions are valid:

1. The edge {$x,y$} is not present.
2. If $X$ is any subset of the game position that contains $x$, and if $x$ is then replaced by $y$, the resulting subset is also a subset of the game position.
3. If $Y$ is any subset of the game position that contains $y$, and if $y$ is then replaced by $x$, the resulting subset is also a subset of the game position.

Then the vertices $x$ and $y$ can be removed, along with all simplexes that contain them, without changing who wins (provided they play the best strategy available). Using this technique, the proof that with optimal play Bob wins Subset Takeaway for $n = 5$, 6 becomes much simpler and takes only a few minutes' analysis.

Back to my question about the 'complement' strategy. When $n = 4$, Alice may start with either a 0-simplex (vertex), a 1-simplex (edge), or a 2-simplex (triangular face). If she chooses a vertex and Bob selects the complement then the game reduces to the case $n = 3$ and Bob wins. If

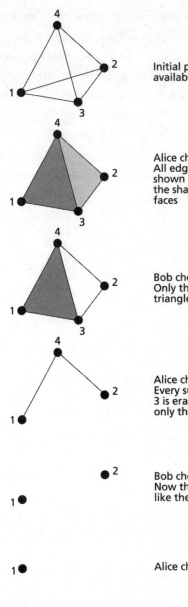

Initial postion. All sub-simplexes available except {1,2,3,4}

Alice chooses {1,2}
All edges and vertices shown available, plus the shaded triangular faces

Bob chooses {2,3,4}
Only the interior of one triangle is erased

Alice chooses {3}
Every subset containing 3 is erased, leaving only this

Bob chooses {4}
Now the game is just like the case n = 2

Alice chooses {2}

Bob chooses {1} and Alice is stuck

**Figure 62** A typical game for a set with four members.

she chooses a triangular face and Bob chooses the complementary point the again the game reduces to the case $n = 3$.

But what if Alice chooses an edge (which by choosing numbers we can assume is {1,2}) and Bob chooses the complementary edge {3,4}? Figure 63 shows that eventually Bob is unable to choose a complementary subset, because the subset concerned is not a simplex. So the 'complementary' strategy fails because it does not specify a legal move. However, with proper strategy, Bob still wins when $n = 4$. Christiansen and Tilford conjecture that for all $n$, Bob's correct response to any

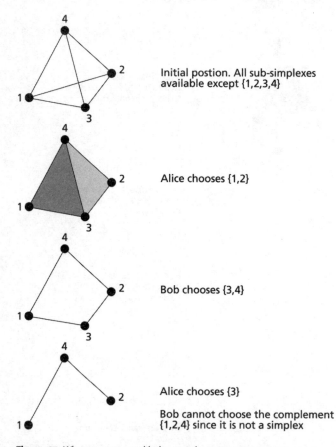

Initial postion. All sub-simplexes available except {1,2,3,4}

Alice chooses {1,2}

Bob chooses {3,4}

Alice chooses {3}

Bob cannot choose the complement {1,2,4} since it is not a simplex

**Figure 63** What goes wrong with the complementary strategy.

opening move by Alice is to choose the complementary subset for his first move. Thereafter, however, he may be forced to deviate from choosing the complement of Alice's move, as we've just seen.

A similar game can be played on any simplicial complex. It might be hoped that whenever the game is played on some triangulation of a simplex (that is, a simplicial complex obtained by subdividing a simplex) then it is a win for Alice (not Bob). Indeed, if this were true it would imply Gale's conjecture (I leave you to work out why, and why it is Alice that we might expect to win). However, Figure 64 shows such a triangulation in which Bob wins – again I leave it to you to work out why.

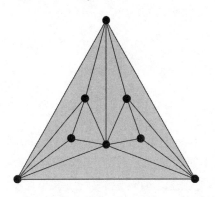

**Figure 64**
Bob wins on this simplicial complex, too.

Computer searches may well prove or disprove Gale's conjecture for $n = 7$, 8, or other small values. For larger $n$, what's required is a new *idea*.

# Easter Is a Quasicrystal

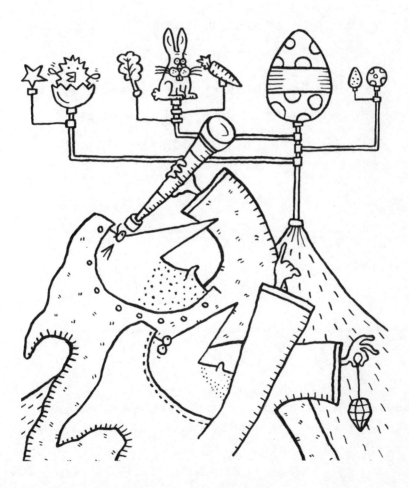

Easter falls on the first Sunday after, but not on the same day as, the first full Moon occurring on, or after, the Spring equinox, which by convention is taken to be 21 March, even when it isn't . . . And it's not the *real* Moon, but an ecclesiastical fiction . . .
Oh, heck, let's just predict the date of Christmas instead.

**M**y first **Mathematical Recreations column** for *Scientific American* was about Fermat's Christmas Theorem. With Easter fast approaching, it seemed only fitting to devote my 96th and final column to Easter. This chapter is based on that final column.

Christmas always falls on 25 December, so there is no difficulty in calculating the date of Christmas . . . but Easter is quite another matter. Easter can fall on any date between 22 March and 25 April, a five-week window. The early Christian Church devised its own methods for calculating the date of Easter.

Mathematicians got in on the act when Carl Friedrich Gauss, generally considered the greatest mathematician ever, invented a simple set of rules that required knowing only the year concerned. Unfortunately, Gauss's work contained a minor oversight, so that it gives 13 April for the year 4200 when the correct date should be 20 April. He corrected this error by hand on his own copy of the published paper.

The first correct purely mathematical procedure was given by an anonymous American in 1876 in the scientific journal *Nature*. In 1965 Thomas H. O'Beirne published two such procedures in his book *Puzzles and Paradoxes*, and I will describe one of them below. More recently the crystallographer Alan MacKay (University College, London) noticed that Easter is a time-quasicrystal – an enigmatic remark that I shall also explain.

The date of Easter changes from year to year for a number of historical reasons. First, the date has to be a Sunday because the crucifixion was on Friday and the resurrection was on Sunday. The timing relative to the Jewish Passover implies that Easter should be closely related to the Passover, which is celebrated for a week following the first full moon of Spring.

The date of Easter was thus linked to several different astronomical cycles, and it is here that the real difficulties arise. The lunar month is currently about 29.53 days long, and the solar year is 365.24 days long. This leads to 12.37 lunar months per year, an inconvenient relationship because it is not an integer. It so happens that 235 lunar months are very close to 19 solar years, and the Church's system for assigning a date to Easter exploits this coincidence.

In 325 CE the Council of Nicaea decided that Easter should fall on the first Sunday after (but not on the same day as) the first full Moon occurring on, or after, the Spring equinox. This is the date in March on which day and night have equal length: they become equal again on the vernal equinox, which occurs in September. Moreover, by convention the Spring equinox would be taken as 21 March. However, this was only one key event in a complex history, as we will see. Leap years could take the true equinox to 22 March on occasion: this possibility was ignored. In those days the year was based on the Julian calendar, with one leap year in every four; full moons were assumed to repeat after exactly 19 Julian years of 365 and 1/4 days. Some juggling with calendar conventions for lunar months made this period equal to 235 lunar months of 29 or 30 days (occasionally 31 in a leap year). The cycle repeated exactly every 76 years – four cycles of 19 years, after which time the pattern of leap years would repeat. The mathematical principle here is that cycles of two different whole number lengths must be repeated a number of times equal to their lowest common multiple before both cycles get back to where they were originally, and 76 is the lowest common multiple of 19 and 4.

This 19-year period was called the lunar cycle, and the year's position in this cycle was indicated by its so-called Golden Number, which ran from 1 to 19 and then repeated starting from 1 again. Easter dates repeated in a 532-year cycle.

It was a tidy system, but unfortunately the mathematics did not accurately respect the true lengths of the lunar month and solar year, and as time passed the calendar started to slip relative to the seasons. (The famous writer Dante Alighieri pointed out that eventually January would cease to be a part of winter.) Discussion continued for over a thousand years, until 1582, when Pope Gregory reformed the civil calendar by omitting leap years in years that were multiples of 100, except that multiples of 400 remained leap years (as the year 2000 was, for example). In order to correct for the previous slippage, ten days were omitted between 4 and 15 October.

Gregory was advised by the astronomer Clavius, and very few relevant phenomena escaped Clavius's attention. In addition to the golden number, the Church's calculating procedure included a second quantity called the *epact*, an integer between 1 and 30 that gives the assumed age of the Moon (starting from $0 = 30 =$ new Moon) in days, immediately before 1 January of the year concerned. At the start of each century, the cycle of epacts is revised, but the cycle of golden numbers continues with no glitches. The choice of epact corrects the errors in the Julian calendar, and also corrects for the fact that 235 lunar months do not exactly equal 19 solar years. Such corrections did not occur in 1900, 2000, or 2001, but one was needed in 2002.

This system is a compromise. The real astronomical equinox can occur as early as 19 March – as will happen in 2096 – or very late on 21 March, as in 1903. In 1845 and 1923 the astronomical full Moon occurred *on* Easter Sunday in most parts of the world, and in easterly longitudes it occurred on Easter Monday. In 1744 there was a full Moon on a Saturday, eight days before Easter Sunday, except that in very westerly longitudes the full Moon occurred on the Friday.

The real Moon does not slavishly follow the ecclesiastical conventions.

To complete its calculations, the church employed a system of letters ABCDEFG for the seven days of the week, starting with A on 1 January. Each year had a *Dominical Letter*, corresponding to which letter was

Sunday. Since all the other calculations ignore 29 February in leap years (it is identified for these purposes with 1 March) there have to be two Dominical Letters in a leap year – one for January and February, the other for the remaining months. Armed with all this information, it is possible to tabulate the relevant aspects of the calendar for any given year, and find the date of Easter.

O'Beirne's method incorporates the various cycles and adjustments into an arithmetical scheme, which I will now indicate and apply to the year 2001.

Let the year of the Gregorian calendar that is under consideration be $x$. Now carry out the following ten calculations (it is easy to program them on a computer):

1. Divide $x$ by 19 to get a quotient (which we ignore) and a remainder $A$.
2. Divide $x$ by 100 to get a quotient $B$ and a remainder $C$.
3. Divide $B$ by 4 to get a quotient $D$ and a remainder $E$.
4. Divide $8B + 13$ by 25 to get a quotient $G$ and a remainder (which we ignore).
5. Divide $19A + B - D - G + 15$ by 30 to get a quotient (which we ignore) and a remainder $H$.
6. Divide $A + 11H$ by 319 to get a quotient $M$ and a remainder (which we ignore).
7. Divide $C$ by 4 to get a quotient $J$ and a remainder $K$.
8. Divide $2E + 2J - K - H + M + 32$ by 7 to get a quotient (which we ignore) and a remainder $L$.
9. Divide $H - M + L + 90$ by 25 to get a quotient $N$ and a remainder (which we ignore).
10. Divide $H - M + L + N + 19$ by 32 to get a quotient (which we ignore) and a remainder $P$.

Then Easter Sunday is the $P$th day of the $N$th month (where 3 = March, 4 = April).

Also: the golden number is $A + 1$, and the epact is whichever of $23 - H$ or $53 - H$ is positive. The Dominical letter can be found by dividing $2E + 2J - K$ by 7 and taking the remainder. Then $0 = A$, $1 = B$, $2 = C$, and so on.

Let's try this method with $x = 2001$. Then (1) $A = 6$; (2) $B = 20$, $C = 1$; (3) $D = 5$, $E = 0$; (4) $G = 6$; (5) $H = 18$; (6) $M = 0$; (7) $J = 0$, $K = 1$; (8) $L = 6$; (9) $N = 4$; (10) $P = 15$. So Easter 2001 was 15 April.

Roughly speaking, the ten steps have the following effects:

1. Find position of year in 19-year cycle. (In fact $A + 1$ is the golden number of the year.)
2. Leap year rule for Gregorian calendar: $B$ increases by 1 for each century year (multiple of 100).
3. $D$ increases only in century years, $E$ gives number of century years that have *not* been leap years.
4. $G$ is month correction to the epact.
5. $H$ is equivalent to the epact (which is $23 - H$ or $53 - H$, whichever is positive.)
6. $M$ deals with an exceptional case regarding the epact. In fact, $M = 0$ unless either $H = 29$ (when $M = 1$ and the epact is 24), or $H = 28$ and $A > 10$ (when again $M = 1$).
7. Start of calculation of day of week for Easter full Moon. Deals with ordinary leap years.
8. Derives full Moon date from epact.
9. Finds month of Easter.
10. Finds day of month for Easter.

In general terms, the date of Easter slips back successively by eight days each year, except that it sometimes increases instead because of various effects (leap years, cycle of the moon, and so on) in a way that looks irregular but actually follows the above arithmetical procedure. Alan MacKay realized that this near-regular slippage ought to show up

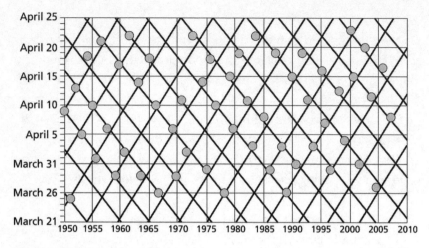

**Figure 65** The Easter quasicrystal from 1950 to 2010.

in a picture of the date of Easter versus the year number (Figure 65). The result is approximately a regular lattice, like the atomic lattice of a crystal (MacKay is a crystallographer). However, the peculiarities of the calendar make the dates vary slightly compared to the lattice, so the diagram is a *quasicrystal*.

Quasicrystals are not as regular as crystals (whose atoms form a precise lattice) but are by no means random. They were discovered in connection with a curious class of tilings of the plane discovered by the Oxford physicist Roger Penrose. In these tilings, two shapes of tile are used, and they tile the plane exactly, but without repeating the same pattern periodically. The atoms of quasicrystals have the same almost-regularity.

With the Gregorian rules in operation, the cycle of Easter dates repeats exactly after 5,700,000 years: these include 70,499,183 lunar months and 2,081,882,250 days. Long before the first repeat, though, the rules will have slipped relative to the astronomical reality. In any case, the lengths of the month and day are slowly changing, mainly because of tidal friction.

Other factors could change things, too. In the United Kingdom, a decision made by Parliament in 1928 makes it possible for the date of Easter to be fixed, without further debate, to be the first Sunday after the second Saturday in April, provided the relevant religious authorities agree. So perhaps in the future the calculation of Easter will be simplified. Until then, though, it is a wonderful example of integer approximations to astronomical cycles, complete with its own intriguing geometrical interpretation. And you can have fun programming the Easter rules and working out, for example, what date Easter will be in the year 1,000,000.

[ANSWER: 16 April, just like 2006.]

# Further Reading

**Chapter 1: Your Half's Bigger than My Half!**

Steven Brams, Alan D. Taylor, and William S. Zwicker, Old and new moving-knife schemes, *Mathematical Intelligencer* vol. 17, no. 4 (1995) 30–5.

Steven Brams, Alan D. Taylor, and William S. Zwicker, A moving-knife solution to the four-person envy-free cake-division problem, *Proceedings of the American Mathematical Society* vol. 125 (1997) 547–54.

Jack Robertson and William Webb, *Cake Cutting Algorithms*, A. K. Peters, Natick, MA 1998.

**Chapter 2: Repealing the Law of Averages**

William Feller, *An Introduction to Probability Theory and Its Applications Volume 1*, Wiley, New York 1957.

**Chapter 3: Arithmetic and Old Lace**

David Gale, *Tracking the Automatic Ant*, Springer, New York 1998.

John H. Halton, The shoelace problem, *Mathematical Intelligencer* vol. 17 (1995) 36–40.

**Chapter 4: Paradox Lost**

David Borwein, Jonathan Borwein, and Pierre Maréchal, Surprise maximization, *American Mathematical Monthly* vol. 107 no. 6 (2000) 517–27.

Jules Richard, Les principes des mathématiques et le problème des ensembles, *Revue Générale des Sciences Pures et Appliquées* (1905); translated in J. van Heijenoort (ed.), *From Frege to Gödel: A Source Book in Mathematical Logic 1879–1931*, Harvard University Press, Cambridge, MA 1967.

J. Richard, Lettre à Monsieur le rédacteur de la revue générale des sciences, *Acta Mathematica* vol. 30 (1906) 295–6.

### Chapter 5: Tight Tins for Round Sardines

Hans Melissen, Packing and covering with circles, Ph.D. thesis, University of Utrecht, 1997.

K. J. Nurmela and P. R. J. Östergård, Packing up to 50 circles inside a square, *Discrete Computational Geometry* vol. 18 (1997) 111–20.

K. J. Nurmela, Minimum-energy point charge configurations on a circular disk, *Journal of Physics A* vol. 31 (1998) 1035–47.

### Chapter 6: The Never-Ending Chess Game

Paul R. Halmos, *Problems for Mathematicians Young and Old*, Dolciani Mathematical Expositions 12, Mathematical Association of America, Washington, DC 1991.

### Chapter 8: Zero Knowledge Protocols

Neal Koblitz, *A Course in Number Theory and Cryptography*, Springer, New York 1994.

### Chapter 9: Empires on the Moon

Joan P. Hutchinson, Coloring ordinary maps, maps of empires, and maps of the Moon, *Mathematics Magazine* vol. 66 (1993) 211–26.

### Chapter 10: Empires and Electronics

Joan P. Hutchinson, Coloring ordinary maps, maps of empires, and maps of the Moon, *Mathematics Magazine* vol. 66 (1993) 211–26.

## Chapter 11: Resurrection Shuffle

Persi Diaconis, Ron Graham, and Bill Kantor, The mathematics of perfect shuffles, *Advances in Applied Mathematics* vol. 4 (1983) 175–96.

Martin Gardner, *Mathematical Carnival*, Penguin and Alfred A. Knopf, New York 1975.


## Chapter 12: Double Bubble, Toil and Trouble

Richard Courant and Herbert Robbins, *What Is Mathematics?* Oxford University Press, Oxford 1969.

Michael Hutchings, Frank Morgan, Manuel Ritoré, and Antonio Ros, Proof of the double bubble conjecture, *Electronic Research Announcements of the American Mathematical Society* vol. 6 (2000) 45–9. Details online at www.ugr.es/~ritore/bubble/bubble.pdf.

Cyril Isenberg, *The Science of Soap Films and Soap Bubbles*, Dover, New York 1992

Frank Morgan, The double bubble conjecture, *Focus* vol. 15 no. 6 (1995) 6–7.

Frank Morgan, Proof of the double bubble conjecture, *American Mathematical Monthly* vol. 108 (2001) 193–205.


## Chapter 13: Crossed Lines in the Brick Factory

Nadine C. Myers, The crossing number of $C_m \times C_n$: a reluctant induction, *Mathematics Magazine* vol. 71 (1998) 350–9.


## Chapter 14: Division without Envy

Steven Brams and Alan D. Taylor, An envy-free cake division protocol, *American Mathematical Monthly* vol. 102 (1995) 9–18.

Steven Brams, Alan D. Taylor, and William S. Zwicker, A moving-knife solution to the four-person envy-free cake-division problem, *Proceedings of the American Mathematical Society* vol. 125 (1997) 547–54.

Jack Robertson and William Webb, *Cake Cutting Algorithms*, A. K. Peters, Natick, MA 1998.

## Chapter 15: Furiously Flashing Fireflies

J. Buck and E. Buck, Synchronous fireflies, *Scientific American* vol. 234 (1976) 74–85.

Renato Mirollo and Steven Strogatz, Synchronisation of pulse-coupled biological oscillators, *SIAM Journal of Applied Mathematics* vol. 50 (1990) 1645–62.

C. Peskin, *Mathematical Aspects of Heart Physiology*, Courant Institute of Mathematical Sciences, New York University, New York 1975, pp. 268–78.

## Chapter 16: Why Phone Cords Get Tangled

Colin Adams, *The Knot Book*, W. H. Freeman, San Francisco 1994.

Richard B. Sinden, *DNA Structure and Function*, Academic Press, San Diego 1994.

## Chapter 17: Sierpinski's Ubiquitous Gasket

Ian Stewart, Le lion, le lama et la laitue, *Pour la Science* vol. 142 (1989) 102–7.

Marta Sved, Divisibility – with visibility, *Mathematical Intelligencer* vol. 10 no. 2 (1988) 56–64.

## Chapter 18: Defend the Roman Empire!

Charles S. ReVelle and Kenneth E. Rosing, Can you protect the Roman Empire? *Johns Hopkins Magazine* (April 1997) 40 (solution on p. 70 of the June 1997 issue).

## Chapter 19: Triangulation Takeaway

J. Daniel Christiansen and Mark Tilford, David Gale's subset takeaway game, *American Mathematical Monthly* vol. 104 (1997) 762–6.

Richard J. Nowakowski (ed.), *Games of No Chance*, Cambridge University Press, Cambridge 2002.

## Chapter 20: Easter Is a Quasicrystal

Alan L. MacKay, A time quasi-crystal, *Modern Physics Letters B* vol. 4 no. 15 (1990) 989–91.

Thomas H. O'Beirne, *Puzzles and Paradoxes*, Oxford University Press, Oxford 1965.

# Index